Genetic Engineering

Second Edition

THE MEDICAL PERSPECTIVES SERIES

Advisors:

B. David Hames *School of Biochemistry and Molecular Biology, University of Leeds, UK.*

David R. Harper *Department of Virology, Medical College of St Bartholomew's Hospital, London, UK.*

Andrew P. Read *Department of Medical Genetics, University of Manchester, Manchester, UK.*

Oncogenes and Tumor Suppressor Genes
Cytokines
The Human Genome
Autoimmunity
Genetic Engineering
Asthma
HIV and AIDS
Human Vaccines and Vaccination
Antibody Therapy
Antimicrobial Drug Action
Molecular Biology of Cancer
Antiviral Therapy
Molecular Virology *Second Edition*
DNA Fingerprinting *Second Edition*
Understanding Gene Therapy
Molecular Diagnosis
Genetic Engineering *Second Edition*

Genetic Engineering

Second Edition

J. Williams[a], A. Ceccarelli[b] and A. Wallace[c]

[a]University of Dundee, Dundee, UK
[b]Dipartmento Scienze Cliniche e Biologîche, Ospedale S. Luigi Gonzaga, Torino, Italy
[c]St Mary's Hospital, Manchester, UK

© BIOS Scientific Publishers Limited, 2001

First published 1993
Reprinted with revisions 1995
Second Edition 2001

A CIP catalogue record for this book is available from the British Library.

ISBN 1 85996 072 3

BIOS Scientific Publishers Ltd
9 Newtec Place, Magdalen Road, Oxford OX4 1RE, UK
Tel. +44 (0)1865 726286. Fax +44 (0)1865 246823
World Wide Web home page: http://www.bios.co.uk/

Published in the United States of America, its dependent territories and Canada by Springer-Verlag New York Inc., 175 Fifth Avenue, New York, NY 10010-7858, in association with BIOS Scientific Publishers Ltd.

Published in Hong Kong, Taiwan, Cambodia, Korea, The Philippines, Brunei, Laos, and Macau only by Springer-Verlag Hong Kong Ltd, Unit 1702, Tower 1, Enterprise Square, 9 Sheung Yuet Road, Kowloon Bay, Kowloon, Hong Kong, in association with BIOS Scientific Publishers Ltd.

Production Editor: Paul Barlass
Typeset by Marksbury Multimedia Ltd, Bath, UK
Printed by The Cromwell Press, Trowbridge, UK

Contents

Abbreviations   **vii**

Preface   **ix**

1   Gene organization and expression   **1**
General principles of gene organization and replication   1
Common features of prokaryotic and eukaryotic gene expression   3
Characteristic features of prokaryotic gene organization and
   expression   9
Eukaryotic mRNA structure and post-transcriptional regulation   11
Eukaryotic gene transcription and its regulation   15
Transcriptional termination and 3′ processing   16
Splicing   17
Chromosome structure   18
References   20
Further reading   20

2   Gene analysis techniques I: low-resolution mapping of genomic DNA   **23**
Nucleic acid hybridization   23
Gel electrophoresis and detection of nucleic acids by Southern
   and Northern blotting   25
Mapping genes to chromosomes   33
Restriction enzyme mapping and its applications   34
References   45
Further reading   45

3   Gene analysis techniques II: determining the structure of genes   **47**
DNA sequence analysis   47
Transcript mapping techniques   55
Searching for genes using computers   58
The polymerase chain reaction   60
References   63
Further reading   64

4   Application of PCR technology to human diseases   **65**
Genotyping using PCR technology   65
Molecular analysis of clinical samples   66
Diagnosing infectious diseases by PCR   76
References   78
Further reading   79

5   Cloning in *E. coli* I: vectors and genomic libraries   **81**
Cloning and cloning vectors   81
Plasmid vectors and their use in manipulating DNA   82
Phage vectors   89
Construction and screening of genomic libraries   96
References   100
Further reading   101

6   Cloning in *E. coli* II: isolating genes   **103**
Construction of cDNA libraries   103
Screening cDNA libraries   106
Identifying genes by positional cloning: cystic fibrosis   114
References   117
Further reading   117

7   Cloning in higher organisms   **119**
Cloning in yeast   119
Transformation of mammalian cells in tissue culture   125
The creation and use of transgenic mice   131
The impact of whole genome DNA sequence determination:
    proteomics and micro-arrays   138
References   140
Further reading   141

Index·   **143**

Abbreviations

AIDS	acquired immune deficiency syndrome
ARMS	amplification refractory mutation detection system
ARS	autonomously replicating sequence
ATP	adenosine triphosphate
β-gal	β-galactosidase
BAC	bacterial artificial chromosome
CCM	chemical cleavage of mismatches
cDNA	copy DNA
CF	cystic fibrosis
CMV	cytomegalovirus
d.p.m.	disintegrations per minute
DCR	domain control region
DM	myotonic dystrophy
DNA	deoxyribonucleic acid
dsDNA	double-stranded DNA
EBV	Epstein–Barr virus
ES	embryonal stem
FISH	fluorescent *in situ* hybridization
HBV	hepatitis B virus
HD	Huntington's disease
hnRNA	heterogeneous nuclear RNA
HV	herpes virus
IREs	iron response elements
LCR	locus control region
Lod	logarithm of the odds
LTR	long terminal repeats
MCS	multi-cloning site
mRNA	messenger RNA
OLA	oligo ligation assay
ORF	open reading frame
PAC	P1-derived artificial chromosomes
PAGE	polyacrylamide gel electrophoresis
PCR	polymerase chain reaction
PFGE	pulsed-field gel electrophoresis
RACE	random amplification of cDNA ends
RNA	ribonucleic acid
RFLP	restriction fragment length polymorphism

rRNA	ribosomal RNA
snRNA	small nuclear RNA
SSCP	single strand conformation polymorphism
ssDNA	single-stranded DNA
tRNA	transfer RNA
VNTR	variable number tandem repeat
YAC	yeast artificial chromosome

Preface

Over the last 20 years there has been a radical change in the way biological problems are investigated and this has resulted in profound new insights into the functioning of biological systems. The key to these advances was the finding that genes in a simple organism, such as the gut bacterium *E. coli*, function in a way fundamentally similar to genes in higher organisms such as man. The nucleus of each human cell contains a thousand times as much DNA as an *E. coli* cell. The relatively vast amounts of DNA present in mammalian cells, and the absence of the powerful genetic methods of analysis that can be applied to *E. coli*, made human genes essentially inaccessible. The crucial realization was that, using bacterial cells as carriers, genes from other higher organisms could be separated, one from another, and amplified to yield large amounts of DNA. From this has flowed an exponentially increasing stream of information that has major implications for medical science. The aim of this book is to describe some of the techniques which have made these advances possible and to show how they are being applied to clinical problems.

Chapter 1

Gene organization and expression

1.1 General principles of gene organization and replication

Genetic information in all living cells is encoded by deoxyribonucleic acid (DNA). DNA is composed of a phosphate-sugar backbone with one of four different bases, adenine, guanine, cytosine or thymine, covalently attached to the sugar residues (*Figure 1.1*). A unit of base + sugar is a nucleoside; the four nucleosides in DNA are adenosine, guanosine, cytidine and thymidine. A unit of base + sugar + phosphate is a nucleotide.

The genome of all living cells is composed of double-stranded DNA that contains two anti-parallel polynucleotide strands. The bases have the potential to form base pairs, joined by non-covalent hydrogen bonds. Adenine forms two hydrogen bonds with thymine, and cytosine forms three hydrogen bonds with guanine. This specificity of base pairing allows the precise duplication of DNA, and enables it to act as the repository of genetic information. The two DNA strands are wrapped around each other to form a double helix. In double-stranded DNA there is one turn of the helix every 3.4 nm (*Figure 1.2*). During DNA replication the helix is unwound and each strand acts as a template for the synthesis of a complementary strand. The incoming nucleotide has a triphosphate group on the 5′ carbon atom of the sugar. This condenses with a hydroxyl group on the 3′ position of the growing chain (*Figure 1.3a*). Thus the chain grows in a 5′ to 3′ direction, and the sequence of DNA is always written with this polarity. This reaction is enzymatically catalyzed by a DNA polymerase. Because DNA chains can grow only in a 5′ to 3′ direction, one of the two strands of the double helix is replicated as a series of discontinuous short pieces (Okazaki fragments), which are subsequently joined up by the enzyme DNA ligase (*Figure 1.3b*).

DNA can be linear or circular. Viruses, unlike cellular organisms, have a variety of genome structures, linear or circular, single-stranded or double-stranded, and DNA or ribonucleic acid (RNA). Some viruses,

1

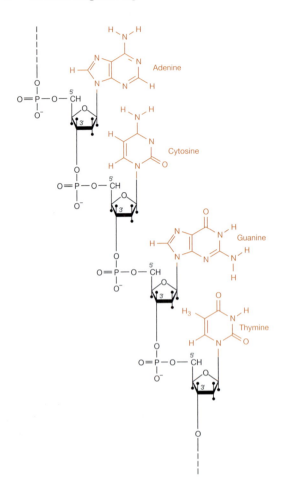

Figure 1.1. A portion of a DNA polynucleotide chain, showing the 3′–5′ phosphodiester linkages that connect the nucleotides.

such as the SV40 DNA tumor virus, have a genome of double-stranded DNA. The genome of a prokaryotic cell, such as *E. coli*, consists of a gigantic circular molecule of DNA (*Table 1.1*).

The DNA molecule found within eukaryotic mitochondria is also circular, although the normal chromosomal DNA is linear. A circular double-stranded DNA molecule in which each strand is intact is termed closed circular DNA (*Figure 1.4a*). If one of the two chains is broken, then its circular shape will still be maintained but the two DNA strands will now be free to rotate around one another. Such a molecule is termed an open circle. At the instant a closed circular DNA molecule is formed from an open circular molecule or a linear molecule of double-stranded DNA, the two ends of the DNA strands are most likely to be rotated relative to one another, very much like a piece of rope thrown into a heap upon the ground. Once the two ends are joined to form a closed circle, this rotation becomes irreversible; the molecule forms a tangled structure and is said to be supercoiled (*Figure 1.4d*). This property of circular DNA is technically

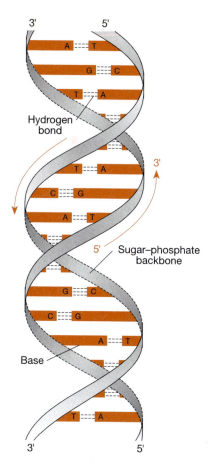

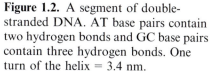

Figure 1.2. A segment of double-stranded DNA. AT base pairs contain two hydrogen bonds and GC base pairs contain three hydrogen bonds. One turn of the helix = 3.4 nm.

important, because open circular and closed circular forms can be separated from one another by virtue of the supercoiled nature of the latter molecule.

1.2 Common features of prokaryotic and eukaryotic gene expression

The genetic information within DNA directs the production of the structural proteins and enzymes which build the cell. The intermediary between DNA and protein is RNA. RNA differs from DNA in containing a ribose, rather than a deoxyribose, sugar and by the substitution of uracil for thymine as the complementary base to adenine. Gene transcription, the copying of the genetic information in DNA to produce an RNA copy, is fundamentally similar to DNA replication, except that only one of the two DNA strands is copied.

There are two quite distinct classes of RNA, messenger RNA (mRNA) and structural RNAs (ribosomal RNA [rRNA] and transfer RNA

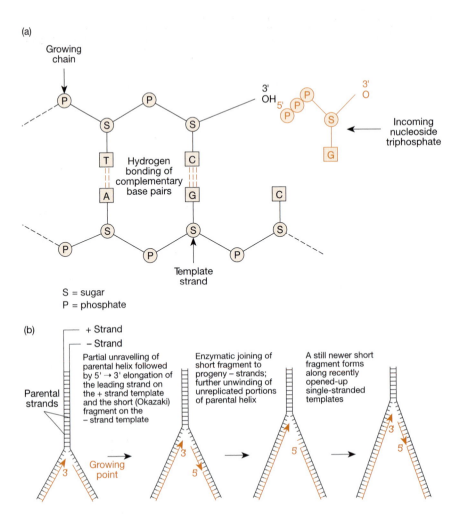

Figure 1.3. The replication of DNA. (a) The chemistry of polynucleotide chain growth. The incoming residue is added to the 3′ end of the growing chain. A condensation reaction forms a covalent bond between the hydroxyl group on the newly synthesized strand and the phosphate group attached to the sugar. The two terminal phosphate groups are liberated. (b) Replication of a double-stranded DNA molecule. At the replication fork the two template strands become unwound one from another. The DNA strand which is extending in the same direction as the replication fork grows in an uninterrupted fashion and is called the leading strand. The lagging strand grows in the opposite direction to the fork by a series of initiation and elongation reactions. At the end of the whole process, the segments of newly replicated DNA on the lagging strand are joined together enzymatically. Reproduced from Watson *et al.* (1987) with permission from The Benjamin Cummings Publishing Company.

Table 1.1. The diversity of DNA organization

Type of DNA	Size (kb)	Topology	Number of encoded genes
Plasmid vector	2–5	Circular	2
E. coli genome	3×10^3	Circular	2000
Human mitochondrial genome	16.5	Circular	38
Typical human chromosome	10^5	Linear	2000

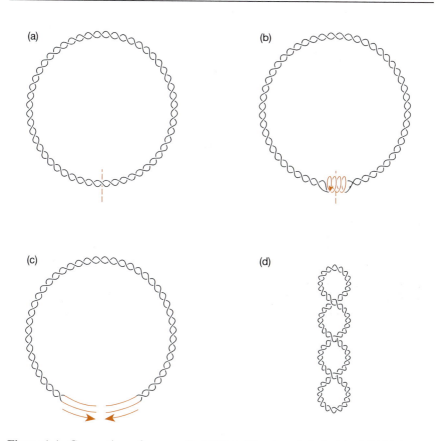

Figure 1.4. Generation of supercoiled DNA. If a closed circle (a) of DNA is cut, rotated in the opposite direction to the twist of the helix (b), and then the cut ends rejoined (c), the DNA becomes negatively supercoiled (d). Reproduced from Watson *et al.* (1987) with permission from The Benjamin Cummings Publishing Company.

[tRNA]) (*Table 1.2*). As its name implies, mRNA is the carrier of genetic information. The typical prokaryotic or eukaryotic cell contains many thousands of different mRNAs, transcribed from different genes and each with the potential to produce a different protein. In both prokaryotic and eukaryotic cells, mRNA forms only a few percent of the total cellular RNA. The structural RNAs, which comprise rRNAs and tRNAs, form the bulk of cellular RNA.

Table 1.2. Classes of RNA molecules

RNA class	Approximate size (bp)	Approximate number of species in a cell	Function	Distribution*
5S rRNA	120	1–2	Ribosomal constituent, large subunit	P, E
5.8S RNA	155	1	Ribosomal constituent, large subunit	E
16S RNA	1600	1	Ribosomal constituent, small subunit	P
23S RNA	3200	1	Ribosomal constituent, large subunit	P
18S RNA	1900	1	Ribosomal constituent, small subunit	E
28S RNA	5000	1	Ribosomal constituent, large subunit	E
tRNA	75–90	100	Translation	P, E
hnRNA	Variable	> 10 000	Precursor of mRNA	E
mRNA	Variable	> 10 000	Encoding polypeptides	E
snRNA	50–230	Tens	RNA processing	E

*E: eukaryotes; P: prokaryotes.

Translation is the process whereby the structural RNAs, and their associated proteins, decode the linear sequence of information contained within the mRNA. The nucleotide sequence of the mRNA is read as a series of triplets (codons) which specify a protein sequence that is co-linear with the mRNA sequence (*Figure 1.5a*). Each codon specifies the insertion of a particular amino acid into the growing peptide chain. Translation begins at an initiation codon, usually AUG, and ends at one of three different termination codons, UAA, UGA or UAG.

There are 20 different amino acids. A language based upon triplets can contain up to 64 possible 'words' (*Table 1.3*). All of the codons are used in the genetic code. Thus there is degeneracy, most amino acids being encoded by more than one codon. Phenylalanine, for example, can be encoded by either a UUU or a UUC codon. The tRNAs act as the intermediaries in translation. They transport a specific amino acid to the mRNA triplet that encodes it. Each tRNA has an anticodon loop containing three nucleotides which is able to pair with one or more of the codons (*Figure 1.5a*). In the example shown, phenylalanine is being incorporated at a UUC codon by a phenylalanine tRNA.

There is no direct recognition by the tRNA molecule of its cognate amino acid. Instead, enzymes called aminoacyl-tRNA synthetases

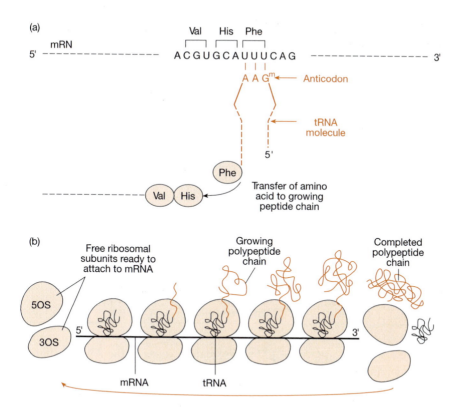

Figure 1.5. The translation of mRNA into protein. (a) Insertion of a single amino acid into the polypeptide chain. This is a highly diagrammatic representation of the process of translation. The codons in the mRNA are recognized by the anticodon loop on the tRNA, in this case a UUC codon is interacting with a phenylalanine tRNA. In all such interactions the last two nucleotides of the anticodon loop in the tRNA are the direct complement of the cognate codon in the mRNA, in this case AA bases in the anticodon loop pair with UU residues in the mRNA. The third residue in the anticodon loop is often a 'non-standard' base, in this case, a methylated guanosine. The base in the third position is called the wobble base because it has the ability to pair with more than one base in the third position of the codon. Thus this phenylalanine tRNA is able to pair with either a UUC or a UUU phenylalanine codon. This enables a cell to utilize fewer tRNAs than would be needed if every codon were recognized by a different tRNA. (b) The structure of a polysome engaged in protein synthesis. The mRNA molecule in this example is moving from left to right.

covalently couple the amino acid encoded by a particular codon to the tRNA that contains the appropriate anticodon. The mRNA is translated upon the ribosomes, bipartite structures that consist of a large and a small subunit (*Figure 1.5b*). Each subunit contains structural RNAs (the rRNAs) complexed with the proteins that perform the various chemical reactions involved in translation. The ribosome moves along the mRNA

Table 1.3. The genetic code

First letter	Second letter				Third letter
	U	C	A	G	
U	UUU (Phe)	UCU (Ser)	UAU (Tyr)	UGU (Cys)	U
	UUC (Phe)	UCC (Ser)	UAC (Tyr)	UGC (Cys)	C
	UUA (Leu)	UCA (Ser)	UAA (STOP)	UGA (STOP)	A
	UUG (Leu)	UCG (Ser)	UAG (STOP)	UGG (Trp)	G
C	CUU (Leu)	CCU (Pro)	CAU (His)	CGU (Arg)	U
	CUC (Leu)	CCC (Pro)	CAC (His)	CGC (Arg)	C
	CUA (Leu)	CCA (Pro)	CAA (Gln)	CGA (Arg)	A
	CUG (Leu)	CCG (Pro)	CAG (Gln)	CGG (Arg)	G
A	AUU (Ile)	ACU (Thr)	AAU (Asn)	AGU (Ser)	U
	AUC (Ile)	ACC (Thr)	AAC (Asn)	AGC (Ser)	C
	AUA (Ile)	ACA (Thr)	AAA (Lys)	AGA (Arg)	A
	AUG (MET)	ACG (Thr)	AAG (Lys)	AGG (Arg)	G
G	GUU (Val)	GCU (Ala)	GAU (Asp)	GGU (Gly)	U
	GUC (Val)	GCC (Ala)	GAC (Asp)	GGC (Gly)	C
	GUA (Val)	GCA (Ala)	GAA (Glu)	GGA (Gly)	A
	GUG (Val)	GCG (Ala)	GAG (Glu)	GGG (Gly)	G

in a 5′ to 3′ direction (*Figure 1.5b*). Translation starts at the AUG initiation codon, which is normally located near to the 5′ end of the mRNA, and the ribosome falls off the mRNA when it reaches a termination codon.

The rate of transit of the ribosome along the mRNA is relatively constant; under optimal conditions it takes about 2.5 s to polymerize 100 amino acids. However, more than one ribosome may be bound to an mRNA at any one time, to form a polyribosome or polysome. The number of ribosomes bound to a particular mRNA determines the rate of its translation: the higher the number of ribosomes bound then the higher will be the number of protein chains produced. Thus the rate of translation of a particular mRNA is generally controlled by the frequency with which translation is initiated.

The protein chain contains the information that directs its post-translational processing and cellular compartmentalization. Soluble cytosolic proteins are released from the polysomes after synthesis. Proteins that are destined to accumulate within the nucleus contain a nuclear localization signal, consisting of a tract of amino acids with basic side chains. Proteins which are destined to be exported from the cell, or to be incorporated into cellular membranes, are transported into the endoplasmic reticulum during their synthesis. Such proteins contain at their N-terminus a stretch of predominantly hydrophobic amino acids, called the signal peptide, which directs the polysomes to the endoplasmic reticulum. There the signal peptide is cleaved off as the growing peptide is secreted into the lumen of the endoplasmic reticulum. The protein is then transported via the Golgi complex to its final destination.

1.3 Characteristic features of prokaryotic gene organization and expression

The primary difference between cells of a eukaryote, such as man, and a prokaryote, such as the bacterium *E. coli*, is the presence or absence of a membrane delimiting the nucleus. In eukaryotes the nuclear membrane acts as a barrier, separating the genetic information contained within the chromosomes (the genome) from the cytoplasm. Medical applications of genetic engineering ultimately concern gene organization and expression in higher eukaryotes, but it is important also to understand the basic features of prokaryotic gene expression because *E. coli* is of such central importance in genetic engineering.

The mechanics of gene expression were first understood in *E. coli*. The genetic information of this bacterium is contained within a circular chromosome of approximately three million base pairs (*Figure 1.6*). Most genes are tightly packed together, with very little superfluous DNA between them. Many genes form part of clusters called operons, containing closely spaced genes which function in a common pathway. The *lac* operon, for example, contains three genes involved in the uptake

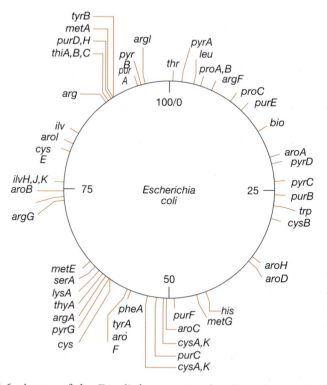

Figure 1.6. A map of the *E. coli* chromosome showing a few of the genetically mapped loci. *E. coli* is by far the best characterized living organism. Its entire genome sequence is known and the function of a large fraction of its genes is understood.

and metabolism of lactose (*Figure 1.7*). Such a cluster is copied into a single mRNA molecule by the enzyme RNA polymerase. Because there is no nuclear membrane to traverse, the ribosomes have access to the mRNA during its synthesis. Hence transcription and translation are essentially co-temporaneous.

Just upstream from the AUG (start) codon of each gene in the operon, the mRNA contains a short tract of sequence known as a ribosome binding site (or sometimes as a Shine–Dalgarno sequence, after its discoverers). A ribosome binding site is essential for efficient translation. It interacts with a sequence in the RNA of the small ribosomal subunit, ensuring that translation commences at the initiation codon.

The enzymes involved in lactose metabolism are produced only when a β-galactoside containing sugar such as lactose is available, and the genes are clustered so that their expression can be co-ordinately induced. In the absence of a β-galactoside, a protein called the lac repressor binds to the DNA just before the start site of transcription of the β-galactosidase gene. This prevents transcription of the gene. β-galactoside-containing sugars bind to the lac repressor, dissociating it from the DNA and allowing the operon to be transcribed. This form of gene regulation is known as transcriptional control. Such control circuits, where a substrate for an enzyme controls synthesis of the enzyme, are very common in prokaryotes.

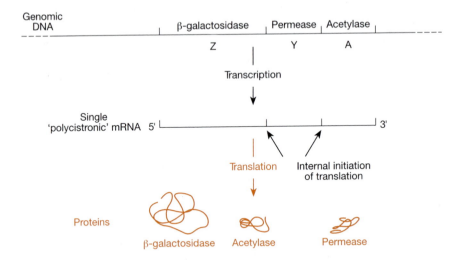

Figure 1.7. The structure and expression of the *lac* operon. This operon contains three genes involved in the metabolism of lactose-containing sugars. It is copied into a single polycistronic mRNA which is translated to give three separate proteins. This mode of expression stands in marked contrast to eukaryotic mRNAs which are monocistronic, and which are generally translated using the AUG nearest the 5′ end. The operator sequences which regulate transcription of the *lac* operon are located just upstream of the *β-gal* gene.

Prokaryotes and eukaryotes use essentially similar mechanisms to decode the information contained within their DNA; however, they differ greatly in cell structure and in the size and organization of their genomes. Not surprisingly, there are also major differences in gene organization and expression which relate to these gross structural differences.

1.4 Eukaryotic mRNA structure and post-transcriptional regulation

The major difference between prokaryotic and eukaryotic gene expression is that, in a eukaryotic cell, mRNAs must pass from the nucleus to the cytoplasm for translation. There is a complex processing pathway for mRNA precursors in the nucleus, and the cytoplasmic mRNA differs in structure from prokaryotic mRNA. In contrast to most prokaryotic mRNAs, each eukaryotic mRNA encodes only a single protein. Hence each mRNA has a single initiation codon and a single termination codon (*Figure 1.8*). After transcription a tract of A residues of approximately 100–200 nucleotides in length is added enzymatically to the 3′ end of the mRNA (*Figure 1.9*). This poly(A) tail has been described as 'God's gift to the molecular biologist', because it allows the quantitation, purification and selective copying of mRNA molecules (see Section 6.1). It acts to protect mRNA from premature degradation once the mRNA has been transported to the cytoplasm and also to facilitate its translation.

Messenger RNA is also blocked at its 5′ end by the post-transcriptional addition of a methylated guanine residue (*Figure 1.10*). This so-called 'cap' facilitates binding of ribosomes to the mRNA during translation and also helps to protect the mRNA from premature degradation. In contrast to prokaryotic mRNA sequences, one mRNA codes for only one protein; the first AUG codon downstream from the cap normally acts to direct

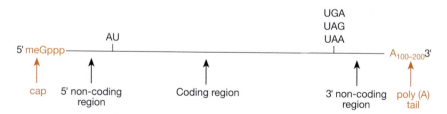

Figure 1.8. The structure of a eukaryotic mRNA. The 'average' mammalian cell contains 10 000–20 000 different mRNA sequences. A typical mRNA is about 2000 nucleotides in length, but there is a very broad range, from a few hundred to greater than 10 000 nucleotides in length. The 5′ non-coding region tends to be only a few hundred nucleotides long, while the 3′ non-coding region can be several thousand nucleotides in length. The mRNA is represented here as a linear structure, but under the conditions found in the cell it would contain many regions of double-stranded RNA, formed by regions of internal homology, and it would be associated with proteins.

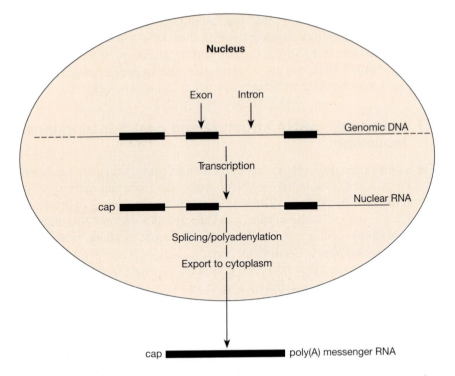

Figure 1.9. Eukaryotic gene expression. This is a highly simplified representation of the path of expression of a gene containing three exons and two introns. It is first transcribed as a nuclear precursor RNA (sometimes called a heterogeneous nuclear RNA (hnRNA) molecule in the older literature), which is longer at the 3′ end than the mature mRNA and which contains introns. Processing enzymes trim away the excess RNA in the nucleus and it is exported to the cytoplasm. During splicing a complex of proteins and small nuclear RNA (snRNA) molecules associate with the splice junctions of the RNA to form a spliceosome. The snRNA molecules contain regions of complementarity to the consensus sequences which lie at the splice junctions and thus act as guides to direct specific cleavage. Polyadenylation occurs after removal of the excess RNA at the 3′ end of the nuclear precursor molecule and is dependent upon the presence of the AAUAAA sequence, which is located just upstream of the poly(A) addition site in most mRNAs.

initiation of translation. While there is no highly conserved ribosome binding site to direct the binding of ribosomes, there are residues just upstream of the initiation codon (often called a Kozak sequence, again after its discoverer [1]), which are weakly conserved between different mRNA sequences and which are necessary for optimal translation.

In general the start site of translation is situated within a few hundred nucleotides downstream of the 5′ end of the mRNA, but the poly(A) tail is often a considerable distance downstream from the translational termination codon. The roles of these non-coding sequences, and the reason for the disparity in length of the 5′ and 3′ non-coding regions, are

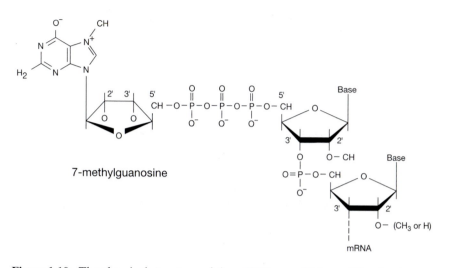

Figure 1.10. The chemical structure of the mRNA cap. This modification is post-transcriptional but the reaction occurs almost co-temporaneously with the initiation of gene transcription.

unclear. However, cytoplasmic mRNA stability and translational efficiency are sometimes regulated by sequences in one or other of these regions. Excellent examples of both of these phenomena are provided by two proteins involved in cellular iron metabolism, ferritin and the transferrin receptor.

Transferrin is a serum glycoprotein that carries iron to the tissues. Upon reaching a target cell, transferrin is bound by the transferrin receptor, which is a protein located in the plasma membrane. The transferrin receptor transports the transferrin molecule with its two associated iron atoms into the interior of the cell. In the 3′ non-coding region of the transferrin receptor mRNA there are nucleotide sequences known as iron response elements (IREs). When iron is abundant within the cell, a protein called IRE-BP binds to the IREs [2,3]. Binding of IRE-BP in some way destabilizes the transferrin receptor mRNA and its concentration thus becomes greatly reduced. IRE sequences have the potential to self-anneal so as to form a stem-loop structure. Such stem-loops are a common feature of regulatory elements in RNA molecules, and the IRE-BP recognizes the apex of the loop formed by the IRE (*Figure 1.11a*).

The above mechanism ensures that at high intracellular iron levels, when there is a reduced need for iron uptake, the number of transferrin receptor molecules in the plasma membrane is maintained at a low level. At the same time the concentration of ferritin, the intracellular iron storage protein, increases. This occurs because of an increase in the rate of translation of its mRNA. The 5′ non-coding region of the ferritin mRNA contains an IRE, and this sequence is responsible for the translational

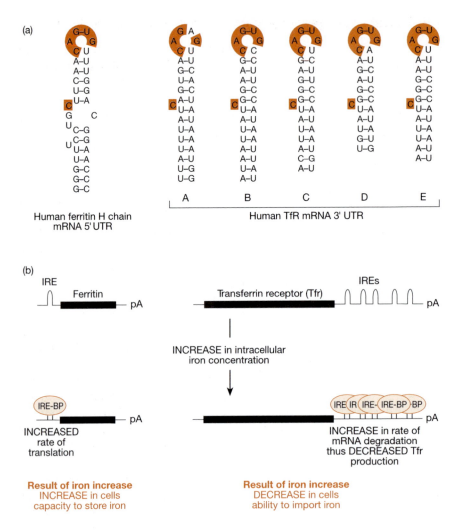

Figure 1.11. Post-transcriptional regulation of gene expression by iron response elements (IREs). (a) Similarities between the IRE in the 5′ non-coding region of the ferritin mRNA and the IREs in the 3′ non-coding region of the transferrin receptor mRNA. (b) Opposing effects of IRE-BP interaction with the transferrin receptor and ferritin mRNAs. This system of post-transcriptional regulation enables the cell to regulate its capacity to transport and store iron, such that it can control the intracellular iron concentration.

regulation, presumably because binding of IRE-BP to the IRE increases the rate of initiation of translation of the ferritin mRNA. Thus, the cell neatly uses the same IRE-BP protein to down-regulate the transferrin receptor and up-regulate ferritin, maintaining the correct balance between them and ensuring that intracellular iron levels are appropriate for the available iron supply (*Figure 1.11*).

1.5 Eukaryotic gene transcription and its regulation

In eukaryotes, there are three separate RNA polymerases which are responsible for the production of different kinds of RNA. Messenger RNA is synthesized by RNA polymerase II, while RNA polymerases I and III synthesize structural RNAs. The earliest processing step in the formation of mRNA is the enzymatic addition of the cap, which occurs almost simultaneously with the initiation of transcription (*Figure 1.9*). Hence the site in the genomic DNA at which transcription starts is commonly known as the cap site.

The interaction of RNA polymerase II with the gene is regulated by a series of proteins known as transcription factors. Transcription factors recognize and bind to relatively short signal sequences in the DNA adjacent to, or in some cases within, the gene. These signal sequences are typically 5–10 nucleotides in length. Generally there is an even smaller core, or consensus, sequence that is conserved between different genes and which is essential for binding.

Close to the cap site in the DNA there are recognition sequences for DNA-binding transcription factors which cause RNA polymerase II to initiate transcription. One very important member of this class of proteins is TFIID. This protein binds to a short AT-rich sequence, the 'TATA box', found approximately 30 nucleotides upstream in a high proportion of eukaryotic genes. The primary function of TFIID is to specify the position at which transcription is initiated. Other DNA-binding proteins, such as Sp1 or CTF, increase the frequency of transcriptional initiation. Thus Sp1 and CTF binding-sites, which are often found upstream of the TATA box, act to increase the overall rate of gene transcription, and are therefore sometimes called efficiency elements.

Post-transcriptional regulation of the kind described above for ferritin and the transferrin receptor may be very important in modulating the level of gene expression, but over-riding control is almost always exerted at the level of gene transcription. Thus globin mRNAs accumulate to such extremely high levels in erythrocytes partly because they become selectively stabilized relative to other transcripts. However, the initial decision to express the globin genes in one cell and not in another is made purely at the transcriptional level; an erythroid cell transcribes the globin genes, but there is no detectable transcription of these genes in a muscle cell.

Cell type-specific gene expression depends on binding of transcription factors to sites which are often known as enhancers. Enhancers may be located upstream of the gene, within an intron, or even beyond the 3' end of the gene. Enhancer-binding proteins can activate gene transcription even when situated thousands of nucleotides away from the cap site. By contrast, transcription control by proteins such as Sp1 or CTF, which bind to efficiency elements, operates only over a short range. Typically, efficiency elements will activate transcription only over a distance of tens or hundreds of nucleotides.

Transcription of a gene displaying a complex pattern of expression may depend on a whole array of binding sites. The metallothionein gene provides a good example. Metallothionein is a small cysteine-rich protein which binds to and detoxifies heavy metals. The human metallothionein MT-IIA gene is expressed at a low level in most tissues and, consistent with this, it has a number of Sp1 sites near the cap site which presumably direct this constitutive expression [4]. Expression of the gene is also inducible by heavy metals, by interferon and by glucocorticoid hormones, and the upstream region contains separate elements responsible for activation by each of these signals (*Figure 1.12*). The generally held model for the expression of such a gene is that distal regulatory and proximal basal transcription factors are brought into contact by looping out of the intervening DNA [5].

Transcription factors contain separate functional domains. A DNA-binding domain tethers the protein to its cognate regulatory element, and an activation domain recruits RNA polymerase II to the 5′ end of the gene. There are several classes of DNA-binding domains which are common to different sorts of transcription factors, and which are conserved between widely divergent groups of organisms. There are also several different kinds of activation domain. Activation domains do not seem to share any particular defined amino acid sequence, but rather are rich in particular amino acids, or even just in particular chemically related classes of amino acids, such as those possessing an acidic side chain. This is in marked contrast to DNA-binding domains, where the order of the bases, and hence the shape of the region that interacts with the DNA, is conserved between members of the same family of transcription factors.

1.6 Transcriptional termination and 3′ processing

In contrast to the good understanding of initiation of gene transcription in eukaryotes, the termination of transcription is less well defined. Transcription proceeds beyond the eventual 3′ end of the mature mRNA,

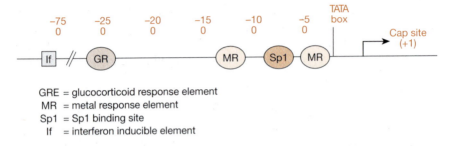

GRE = glucocorticoid response element
MR = metal response element
Sp1 = Sp1 binding site
If   = interferon inducible element

Figure 1.12. Modular structure of the human metallothionein MT-IIA gene. The gene contains DNA sequence elements that are responsible for basal transcription in all tissues, e.g. Sp1 binding sites, and other elements responsible for regulation by various inducing agents.

and the resultant primary transcript is then cleaved internally to generate the mRNA precursor. The portion of the transcript downstream of the cleavage point is degraded within the nucleus. Cleavage occurs 10–20 nucleotides downstream of an AU-rich sequence, AAUAAA, which is highly conserved in all eukaryotic mRNAs. An enzyme called poly(A) polymerase then synthesizes the poly(A) tail at the 3′ terminus of the mRNA.

1.7 Splicing

Most genes in higher eukaryotes are interrupted by non-coding DNA segments. The genes are a patchwork of coding sequences or exons ('exons exit the nucleus' is a useful mnemonic), interspersed with non-coding sequences (introns). Introns are removed from mRNA precursors before they leave the nucleus (*Figure 1.9*). The reaction in which the RNA transcribed from introns is removed is called splicing. The signals which direct splicing flank the intron. Almost all introns have a GU dinucleotide at their 5′ boundary, and an AG dinucleotide at their 3′ boundary. These dinucleotides form part of a larger consensus sequence that overlaps the intron–exon boundaries. Most genes in higher eukaryotes are interrupted by one or more introns which are generally longer than the exons. Hence the major part of the nuclear precursor is removed in order to generate a functional mRNA.

Why do eukaryotes use such an apparently wasteful system? First, it adds an element of flexibility not available to prokaryotes. In some genes the combination of introns which are removed by splicing varies between cell types, allowing a single gene to produce transcripts which differ in primary structure and therefore in coding potential. This form of post-transcriptional regulation is called differential splicing. A good example is provided by the fibronectin gene. Fibronectin is an extracellular protein that mediates the interaction between the cell and extracellular matrix components, such as collagen and heparin. Fibronectin comprises a linear array of repeated amino acid sequences, the different sorts of repeat being responsible for mediating interaction with different matrix molecules. Differential splicing generates multiple forms of fibronectin which differ in their precise biological functions. For example, the fibronectin produced by hepatocytes and secreted into the plasma lacks sequences derived from repeats EIIIA and EIIIB because these are spliced out and degraded in the nucleus (*Figure 1.13*).

Differential splicing is a very economical way for an organism to generate many cell type-specific variants of a protein using a single gene. There is an additional, more general, reason for the existence of splicing. Analysis of the functional domains in proteins shows that they are often encoded by single exons. This is true of fibronectin, for example. Over the course of evolutionary time, genes seem to interchange functional domains by exchanging exons or groups of exons. This 'mix

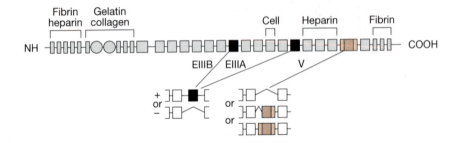

Figure 1.13. Alternative splicing of the fibronectin mRNA. This is a schematic representation of the rat fibronectin gene showing the three positions at which alternative splicing is known to occur. Hepatocytes produce several forms of fibronectin found in plasma. Unlike fibronectins produced by other cells, hepatocyte fibronectins do not include the EIIIA and EIIIB repeats. These are spliced out of the primary transcript in hepatocytes and degraded in the nucleus. The plasma forms differ from one another in the precise pattern of splicing in the V segment. Differences in splicing in the V region also occur between different non-hepatocyte cell types. Reproduced from Ruoslahti (1988) with permission from the *Annual Review of Biochemistry*, Vol. 57, 1988 by Annual Reviews Inc.

and match' arrangement would be much more difficult were it not for the existence of introns, which allow rearrangements to occur within non-functional regions.

These two factors may explain why, in contrast to the parsimony of bacteria, eukaryotes seem so profligate in their use of DNA. The nucleus of every human cell contains about 7 pg of DNA, while an *E. coli* cell contains about 0.003 pg. Some of this extra DNA obviously encodes information necessary to make a man rather than a bacterium, but much of the extra DNA would appear to be totally superfluous. Maybe, however, the introns do have a long-term evolutionary value as internal spacers that facilitate the genetic rearrangements discussed above.

1.8 Chromosome structure

The nucleus of a single diploid human cell contains about 6×10^9 bp of DNA. If stretched to its full length, this DNA would be about 2 m long, but it is contained within a nucleus only 10 μm in diameter. This enormous degree of packaging is achieved by wrapping up the DNA with proteins called histones. In vertebrates there are five histones, H1, H2A, H2B, H3 and H4. The basic packaging unit, or nucleosome, is an octamer composed of two molecules, each of the core histones, H2A, H2B, H3 and H4, forming a disc-shaped structure. Exactly 146 bp of DNA are wound around the core, like a thread on a spool, making slightly less than two complete turns (*Figure 1.14a*). The gap between neighboring nucleosomes is approximately 50 bp in length, and one molecule of histone H1 binds in this linker region. In transcriptionally inactive chromatin there is a further

(a)

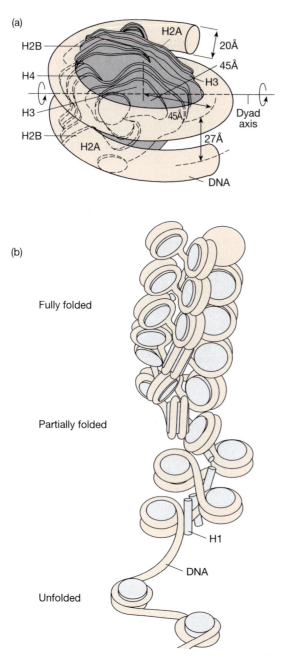

(b)

Fully folded

Partially folded

Unfolded

Figure 1.14. Chromatin structure. (a) The nucleosome. Schematic representation of double-stranded DNA assembled into a nucleosome core particle. The DNA shown as a tube is wrapped around the octamer of core histones. Redrawn from R. D. Kornberg and A. Klug (1981) *Scientific American*, **244**, 52 with permission. (b) The solenoid. The nucleosome core particles are assembled with histone H1 to form the 30 nm solenoid structure. Redrawn from M. Singer and P. Berg (1991) *Genes and Genomes* with permission from University Science Books.

order of packaging to form a structure known as the solenoid (*Figure 1.14b*), comprising nucleosomes wrapped around a multimeric rod of H1 subunits. The solenoid is 30 nm in diameter and each turn contains six nucleosomes and six H1 molecules.

Chromosome condensation may play a role in repressing gene expression, perhaps by excluding components of the transcriptional machinery. It is possible to investigate the structure of chromatin by determining the sensitivity of DNA to endonucleases, enzymes which break the DNA chain. Packaging into higher order structures, such as the solenoid, protects the DNA from cleavage. Genes which are active in a particular tissue are susceptible to nuclease cleavage, while genes which are inactive in that tissue are relatively resistant. Another line of evidence comes from insect salivary glands. These contain giant chromosomes which can be seen under the microscope to be expanded into puffs at the sites of actively expressed genes. Such observations suggest that transcriptionally active chromatin is organized in extended loops where the chromatin is decondensed and available for transcription.

Such a loop may contain a single gene or it may contain a whole cluster of genes that are potentially active in that tissue. One particularly interesting and well-defined example of the latter phenomenon is provided by the human β-like globin gene locus. This spans about 100 000 bp and contains the entire family of β-like globin genes, all of which lie in the same relative transcriptional orientation. The order of the genes in the locus mirrors their temporal pattern of expression, with the embryonic ε-globin gene at the 5′ end of the cluster, the two fetal γ-globin genes in the center, and the adult δ- and β-globin genes at the 3′ end. DNA sequence elements adjacent to the individual genes direct their specific temporal patterns of expression. DNA sequence elements at the extreme 5′ boundary of the locus, in the locus control region (LCR) or domain control region (DCR), are responsible for activation [6], and presumably decondensation, of the locus in erythroid cells. In erythroid cells, over 120 kb of DNA including and surrounding the β-globin cluster is decondensed (evidence for this is shown by the fact that it is hypersensitive to DNAase I and replicates early in the S-phase of the cell cycle), whereas in a non-expressing tissue, such as sperm, it is highly condensed [7]. However, not all transcriptionally active chromatin segments adopt an open or closed configuration as a whole: the α-globin cluster includes segments that are open even in non-expressing cells [8].

References

1. **Kozak, M.** (1986) Point mutations define a sequence flanking the AUG initiator codon that modulates translation by eukaryotic ribosomes. *Cell* **44**: 283.
2. **Klausner, R.D. and Harford, J.B.** (1989) *cis-trans* models for post-transcriptional gene regulation. *Science* **246**: 870.
3. **Casey, J.L., Hentze, M.W., Koeller, D.M., Wright Caughman, S.W., Roault, T.A., Klausner, R.D. and Harford, J.B.** (1988) Iron-responsive elements: regulatory RNA sequences that control mRNA levels and translation. *Science* **240**: 924.
4. **Hamer, D.H.** (1986). Metallothionein. *Annu. Rev. Biochem.* **55**: 913.
5. **Ptashne, M.** (1986) Gene regulation by proteins acting nearby and at a distance. *Nature* **322**: 697.
6. **Grosveld, F., Blom van Assendelft, B., Greaves, D.R. and Kollias, G.** (1987) Position-independent, high-level expression of the human beta-globin gene in transgenic mice. *Cell* **51**: 975.

7. **Townes, T.M. and Behringer, R.R.** (1990) Human globin locus activation region (LAR): role in temporal control. *Trends Genet* **6**: 219.
8. **Vyas, P., Vickers, M.A., Simmons, D.L., Ayyub, H., Craddock, C.F. and Higgs, D.R.** (1992) *Cis*-acting sequences regulating expression of the human alpha-globin cluster lie within constitutively open chromatin. *Cell* **69**: 781.

Further reading

Darnell, J., Lodish, H. and Baltimore, D. (1990) *Molecular Cell Biology*, 2nd edn. Scientific American Books, New York.

Lewin, B. (1990) *Genes IV*. Oxford University Press, Oxford.

Ruoslahti, E. (1988) Fibronectin and its receptors. *Annu. Rev. Biochem.* **57**: 375–413.

Smith, C.W.J., James, J.G. and Nadal-Ginard, B.N. (1989) Alternative splicing in the control of gene expression. *Annu. Rev. Genet.* **23**: 527–577.

Strachan, T. (1992) *The Human Genome*. BIOS Scientific Publishers, Oxford.

Watson, J.D., Hopkins, N.H., Roberts, J.W., Steitz, J.A. and Weiner, A.M. (1987) *Molecular Biology of the Gene*. Benjamin Cummings, Menlo Park, CA.

Chapter 2

Gene analysis techniques I: low-resolution mapping of genomic DNA

Underpinning all the techniques used to isolate and manipulate genes are a series of methods for analyzing and characterizing DNA and RNA sequences. These are described in this and the following chapter. In this chapter we describe the classical methods of nucleic acid hybridization, their application in Southern and Northern blot analyses, and their extension to mapping DNA sequences to particular chromosomal locations and to diagnosis of human genetic diseases.

Gene structure is determined using a combination of DNA and RNA mapping techniques. Most of these methods rely to a greater or lesser extent upon the ability to perform nucleic acid hybridization.

2.1 Nucleic acid hybridization

If a double-stranded DNA molecule is exposed to high temperature, or to very alkaline conditions, then the two strands will come apart. The molecule is said to have become denatured. The temperature at which denaturation occurs is termed the melting temperature or T_m. If the denatured DNA is returned to a temperature below its T_m, or to neutral pH if alkali was used to denature it, each strand will, after a time, find its complementary strand. The two strands will 'zipper' back together to re-form a double-stranded DNA molecule. This ability of complementary sequences to anneal, or hybridize, to one another can be used to identify molecules which contain the same sequence of nucleotides. Hybridization works even when the complementary sequences form part of a complex mixture of nucleic acid molecules.

When employed analytically, hybridization is normally performed using one labeled sequence, termed the probe, and an unlabeled sequence called the target. The probe is labeled by incorporation either of radioactively labeled nucleotides or of nucleotides which are chemically modified so that they can be identified immunologically. The probe is the known, pure species in the hybridization and the target is the unknown species to be identified. The target will most often form part of a mixture of unrelated nucleic acid sequences. The probe is usually added in considerable excess over the target, and its concentration determines the rate of reaction. Hybridization reactions occur with logarithmic kinetics, so that most of the annealing occurs early during the incubation. However, extended hybridization times, up to one or two days, are often used to ensure complete reaction. The target and probe can be hybridized to each other in solution, or alternatively the target can be immobilized on an inert support such as nitrocellulose (*Figure 2.1*). When the target is immobilized the rate of annealing for a given amount of target and probe is greatly reduced relative to that observed when both partners are free to diffuse in solution.

The rate of hybridization depends upon the DNA concentration, the concentration of salt in the solution and the temperature. Hybridization reactions are normally performed at 10–20°C below the T_m of the duplex. Under any given conditions of ionic strength and temperature, the T_m of a duplex depends on the base composition of the annealed partners. AT base pairs form only two inter-strand hydrogen bonds, while GC base pairs form three hydrogen bonds. Hence, the higher the content of GC base pairs within a duplex, the higher will be its T_m. These relationships are described by the following formula for a DNA:DNA duplex:

$$T_m = 16.5 \, (\log \mathrm{Na}^+) + 0.41 \, (\%GC) + 81.5°C$$

where Na^+ is the sodium ion concentration and (% GC) is the percentage of GC residues in the duplex.

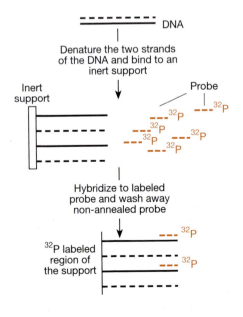

Figure 2.1. The principle of hybridization using an inert support. The target DNA is most often localized to one part of the support and, in most cases, the aim of the hybridization experiment is to identify this region. The probe is prelabeled in some way, most often by the incorporation of radioactivity. After hybridization, the annealed probe is detected by autoradiography.

This relationship holds only for DNA:DNA duplexes. Duplexes in which RNA is one of the annealed partners melt at a higher temperature. DNA:DNA duplexes are less stable (lower T_m) than DNA:RNA duplexes, which in turn are less stable than RNA:RNA duplexes. At high temperatures (e.g. 70°C) RNA is degraded, so hybridizations involving RNA are usually performed in the presence of formamide. Formamide disrupts hydrogen bonding and thus reduces the effective T_m of the duplex, allowing annealing to occur at a lower temperature (e.g. 37°C).

The target and the probe need not be identical over their whole lengths. Non-identical sequences can cross-hybridize if they are sufficiently closely related. The effect of non-complementary base pairs, or mismatches, is to reduce the T_m. So, by controlling the temperature and concentrations of salt and formamide, it is possible to favor or disfavor cross-hybridization. The combination of these three factors is used to dictate the stringency of the hybridization reaction. For sequences shorter than 100–200 nucleotides, the T_m also depends upon the length of overlap between the two sequences, and the effect of mismatches becomes greater the smaller the overlap.

Short oligonucleotides (15–50 nucleotides long) are very frequently used as probes. By controlling the stringency of the reaction, it is possible to control the hybridization of such oligonucleotides to their target with exquisite sensitivity. For short regions of complementarity, such as are obtained using oligonucleotide probes, the mathematical relationship given above breaks down. Here the '2 + 4' rule is used; at an ionic strength of 1.08 M (the ionic strength of 6 SSC, a standard buffer used in nucleic acid hybridization) each AT base pair contributes 2°C of stabilization to the duplex and each GC base pair 4°C. Hence a nucleotide sequence containing 10 AT and 10 GC base pairs has a T_m of 60°C.

2.2 Gel electrophoresis and detection of nucleic acids by Southern and Northern blotting

2.2.1 Gel electrophoresis

Nucleic acid molecules can be distinguished by differences in their chain length, and many of the methods used to study gene organization and expression combine nucleic acid hybridization with a chain length measurement. Because they have a net negative charge at neutral pH, DNA and RNA molecules migrate towards the positive terminal when placed in an electric field (*Figure 2.2*). This process, of electrophoresis, is performed in an agarose or polyacrylamide gel. Nucleic acids are loaded into slots in the gel and allowed to migrate towards the positive terminal. The pores in the gel act to sieve the molecules, so that the mobility of a particular nucleic acid species depends on its length. All the molecules of a particular size move at approximately the same rate through the gel, forming a band that gradually increases in width during electrophoresis

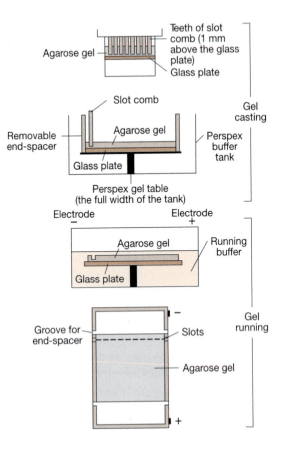

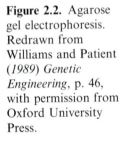

Figure 2.2. Agarose gel electrophoresis. Redrawn from Williams and Patient (*1989*) *Genetic Engineering*, p. 46, with permission from Oxford University Press.

because of diffusion. The size of the molecules in any band upon the gel can be deduced by running DNA molecules of known size in a parallel track on the gel and determining their migration position relative to the species of unknown size. Polyacrylamide has a smaller average pore size than agarose and so is effective at separating molecules of 10–1000 nucleotides in length (1000 nucleotides is termed 1 kilobase or 1 kb), with very high resolution. The larger pores in agarose allow it to resolve much bigger molecules, of up to 100 kb in length. However, the separated molecules migrate as a wider band so that the resolution (i.e. the ability to separate molecules which differ only slightly in size) is lower with agarose than with polyacrylamide. In pulsed-field gel electrophoresis (PFGE) the electrical field across the gel is repeatedly switched back and forth between two or more directions. Since very large molecules change direction less rapidly as they move through the agarose gel, this technique allows extremely long DNA molecules, of many millions of base pairs in length, to be separated. After gel electrophoresis, the nucleic acid molecules must be detected. They can be visualized directly by including ethidium bromide in the polyacrylamide or agarose gel. This dye binds to double-stranded nucleic acids, by intercalating between adjacent base pairs, and emits a red fluorescence when exposed to UV radiation. As little as 5–10 ng of DNA can be detected (*Figure 2.3*). If the nucleic acid molecules are radioactively labeled, then individual resolved species can be detected by autoradiography. Here the sensitivity is dictated largely by one's patience in waiting for the autoradiogram to expose, but an amount of ^{32}P which gives only a few disintegrations per minute (d.p.m.) within a band can readily be detected.

2.2.2 Southern and Northern blotting

Sometimes a complex mixture of unlabeled molecules will be subjected to electrophoresis with the aim of determining the position of migration of just one or a few molecular species. This can be achieved provided that there is a hybridization probe that recognizes the specific molecule(s) of interest. In Southern blotting [1], DNA fragments are separated by

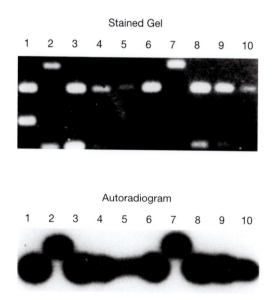

Figure 2.3. Analysis of DNA organization using gel electrophoresis. A DNA molecule was fragmented using 10 different restriction enzymes and the digestion products were subjected to electrophoresis through an agarose gel containing ethidium bromide. The gel was placed on a strong source of UV light and the red fluorescence of the ethidium bromide intercalated into the DNA was recorded photographically. The gel was then subjected to Southern blotting (*Figure 2.4*). The DNA molecule contains an actin gene and the subset of restriction fragments which contain actin DNA sequences was detected using an actin hybridization probe. The identity of the hybridizing bands can be deduced by comparing the two panels.

agarose gel electrophoresis and the double-stranded DNA denatured by incubating the gel in alkali. After neutralization of the alkali, the gel is placed in contact with a reservoir containing a buffered salt solution and overlaid with a filter which is in turn overlaid with dry paper towels (*Figure 2.4*).

Buffer from the reservoir flows through the gel, carrying with it the DNA fragments. Single-stranded nucleic acids bind irreversibly to nitrocellulose filters or can be cross-linked to other, less fragile, supports (such as nylon membranes) by exposure to UV light. During the transfer, DNA fragments retain their positions relative to one another with little loss of resolution. Northern blotting [2] is a similar process that is applied to RNA molecules. Electrophoresis is performed in the presence of a denaturing agent to prevent inter- and intra-molecular interactions that would prejudice separation of the different RNA species.

The membrane or filter bearing the nucleic acids is exposed to a complementary probe under appropriate hybridization conditions. Probes can be either cloned DNA or short chemically synthesized oligonucleotides. Such oligonucleotide probes, 20–50 nucleotides in length, contain the complement of the precise piece of nucleic acid which is to be detected. Chemical DNA synthesis is performed by coupling the 'first' base, in this case the base at the 3′ end of the desired sequence, to an inert resin and

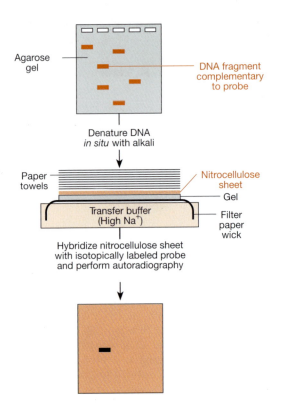

Figure 2.4. Southern blotting.

performing serial cycles of base addition. In contrast to nucleic acid polymerization reactions in nature, the chain grows in a 3′ to 5′ direction. The synthesis is automated using a machine which can synthesize chains up to 100 nucleotides long in a matter of hours. (In other applications of the method, whole genes 1000 or more nucleotides in length have been constructed by joining together overlapping, chemically synthesized chains.) The probe is labeled, prior to hybridization, by incorporation of nucleotides that are either radiolabeled or are chemically modified so as to be detectable immunologically.

2.2.3 Preparation of labeled probes

For synthetic oligonucleotides, a ^{32}P isotopic label can be added to the 5′ hydroxyl group using polynucleotide kinase (*Table 2.1*). Alternatively, labeled nucleotides can be incorporated at the 3′ end using terminal transferase. Because label is added at only one place within the oligonucleotide, such end-labeled probes contain relatively little radio-activity per unit weight of DNA. This low specific activity limits their sensitivity as probes.

If a high specific activity probe is required, then some method of uniform labeling must be used. Here the probe sequence is copied along part or all of its length using DNA polymerase and labeled nucleoside triphosphates. DNA polymerase works by extending a short double-stranded region made by annealing an oligonucleotide primer to the

Table 2.1. Methods of *in vitro* labeling of nucleic acids for Southern and Northern transfer

Labeling method	Enzyme used	Template	Product	Structure at end of reaction*	Specific activity
5′ end labeling	Polynucleotide kinase	DNA or RNA	DNA or RNA	ds or ss	Low
3′ end labeling	Large (Klenow) subfragment of *E. coli* DNA polymerase-1	DNA	DNA	ds	Low
3′ end labeling	Terminal transferase	DNA or RNA	DNA or RNA	ss	Low
Nick translation	*E. coli* DNA polymerase-1	DNA	DNA	ds	High
Random priming	Klenow subfragment of DNA polymerase-1	DNA or RNA	DNA	ds	High
In vitro transcription	Bacteriophage (e.g. T7) RNA polymerase	DNA	RNA	ss	High
PCR (see Chapter 3)	*Taq* polymerase	DNA	DNA	ds	High

*ds: double stranded; ss: single stranded.

single-stranded template. Thus this method of uniform labeling requires a primer which matches the probe sequence. If the probe sequence is not known (as is often the case when natural cloned DNA is used), random oligonucleotide labeling [3] can be used. A large excess of short oligonucleotides, hexamers for example, of random DNA sequence is used as primer (*Table 2.1, Figure 2.5*). These are made by adding a mixture of all four bases at each step in the chemical synthesis reaction. The DNA is denatured and the two complementary strands are copied in the presence of labeled, or chemically modified, nucleoside triphosphates. The polymerase used is a proteolytic fragment (Klenow fragment) derived from DNA polymerase-1 of *E. coli*. Chance homology ensures that the random oligonucleotides anneal to the separated DNA strands at many points along their length, so providing the primers that the polymerase needs for the initiation of DNA synthesis. This is only one of several uniform labeling methods (*Table 2.1*), one of which involves an initial genetic engineering step (see Section 4.2.4).

If the probe is double-stranded DNA, a fraction of the probe will become unavailable for hybridization to the target because the two strands of the probe will anneal together. However, there will normally be a sufficient excess of probe to allow some of the probe to hybridize to the target. At the end of the hybridization, unbound probe is removed by washing. Isotopically labeled probe, bound to its target, is then detected by autoradiography (*Figure 2.6* and *Table 2.2*). Using such methods, tiny amounts of DNA, in the order of 1 pg per band, can be detected. Until relatively recently, radioactive labeling of the probe was the most sensitive detection technique available; now, however, immunodetection procedures, based upon amplification of the signal with enzymes bound to

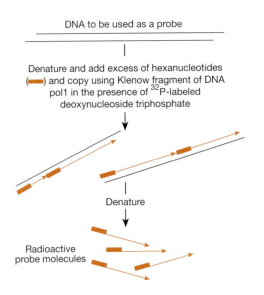

DNA to be used as a probe

Denature and add excess of hexanucleotides
(━━) and copy using Klenow fragment of DNA
pol1 in the presence of ^{32}P-labeled
deoxynucleoside triphosphate

Denature

Radioactive
probe molecules

Figure 2.5. The principle of random primed (oligo-) labeling. The DNA to be used as a probe is denatured by heating and mixed with hexanucleotides of random sequence which then act as primers for DNA synthesis.

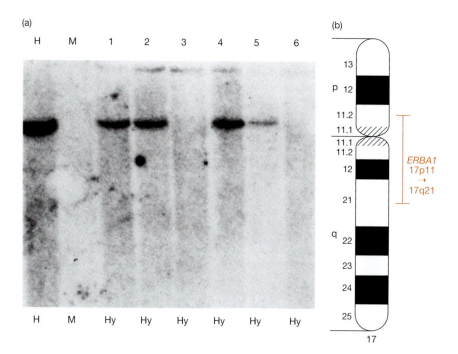

Figure 2.6. Southern blot of a mouse–human hybrid panel. (a) Human–rodent hybrids containing different combinations of human chromosomes can be used to map genes and proteins to specific chromosomes. Because human chromosomes can be lost during continuing culture, the hybrids must be karyotypically characterized every time a batch of DNA is prepared. In this experiment DNA from such a panel was subjected to restriction enzyme digestion and Southern blotting using a probe containing the human *ERBA1* gene. Lanes H and M contain control human and mouse DNA, respectively. Lanes 1–6 contain DNA derived from six different mouse–human hybrids. These results show that four of the hybrids (1, 2, 4 and 5) contain the human *ERBA1* gene and two (3 and 6) do not. Further analysis (*Table 2.2*) of a larger panel of hybrids, one of which contained a translocation of a part of human chromosome 17, allowed mapping of the *ERBA1* gene to the centromere proximal region of chromosome 17. (b) Representation of human chromosome 17 showing location of the *ERBA1* gene as deduced from the mapping information in (a).

Table 2.2. Hybridization of DNA from human-mouse hybrids with *ERBA-1-*specific probes

Hybrids	Result	1	2	3	4	5	6	7	8	9	10	11	12	13	14	15	16	17	18	19	20	21	22	X
DUR4.3	+	−	−	+	−	+	−	−	−	−	+	+	+	+	+	+	−	+	+	−	+	+	+	+
CTP34B4	+	+	+	+	−	+	+	+	+	+	−	−	+	−	+	−	+	+	+	−	−	−	−	+
3W4C15	+	−	−	−	−	−	−	+	−	−	+	+	+	−	+	+	−	+	−	−	−	+	−	+
CTP41P1	+	−	−	+	−	−	+	+	−	−	−	−	−	−	+	−	+	+	−	−	+	−	−	−
PotB2/B2	+	−	−	−	−	−	−	−	−	−	−	−	+	−	−	−	−	+	−	−	−	+	+	−
PCTB/A1.8	+	−	−	−	−	−	−	−	−	−	−	−	−	−	−	−	−	+	−	−	−	−	−	−
Horp27RC14	−	−	−	−	+	−	−	+	−	−	+	+	+	−	+	+	−	−	−	−	−	+	−	+

antibodies, are claimed to be equally sensitive (*Figure 2.7*). The great advantage of non-radioactive probes is that they can be stored for long periods of time because they are not subject to radioactive decay and they require no special safety precautions. The end product of the Southern and Northern transfer is an image of the original gel in which only a subset of the products is visualized (*Figure 2.3*). The position of migration can be used to deduce the size of the RNA or DNA molecules hybridizing

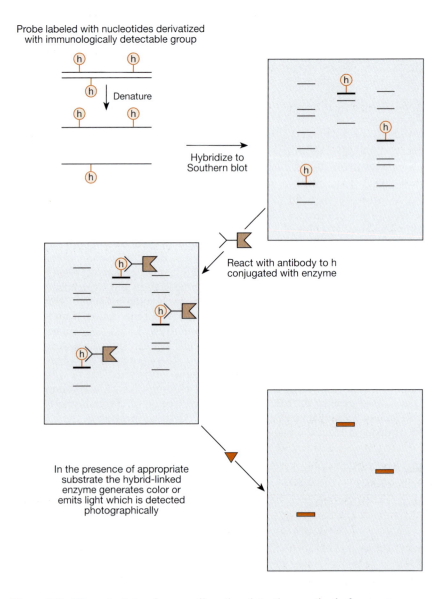

Figure 2.7. The principle of non-radioactive detection methods for *in situ* hybridization.

to the probe, and the strength of the signal is proportional to the amount of nucleic acid within a particular band. Therefore, by analyzing RNA or DNA samples extracted from different sources in adjacent slots, a quantitative comparison can be made of the amount of a specific target molecule in each sample.

2.3 Mapping genes to chromosomes

One of the most important applications of nucleic acid hybridization is in the construction of genetic maps, that is in determining the chromosomal location of individual genes. There are three commonly used methods that differ in the degree of resolution they afford.

2.3.1 Somatic cell hybrids

Somatic cell hybrids are prepared by fusing together human and rodent cells using reagents, such as polyethylene glycol or Sendai virus, which cause membrane fusion. The resulting hybrid cells are then grown in culture medium containing selective agents which allow only the hybrid cells to grow. In hybrids produced using human and rodent cells, the rodent chromosomes are retained, while the human chromosomes are lost. This random chromosome loss occurs during continued culture and shows no apparent selectivity for individual human chromosomes. Thus it is possible to produce a panel of hybrid cells containing different combinations of human chromosomes. Genomic DNA is prepared from panels of such hybrids and used in hybridization analysis (normally a Southern transfer) with the human probe which is to be mapped. By determining which of the panel of hybrids shows a hybridization signal, and correlating this information with the karyotype of the cells, it is possible to deduce which chromosome carries the test gene (*Figure 2.6, Table 2.2*).

2.3.2 In situ *hybridization*

In this technique a labeled DNA probe is hybridized to metaphase chromosomes and the position of hybridization is visualized at the light microscope level. Metaphase chromosomes are liberated from cells and bound to the surface of a cover slip. The DNA within the chromosomes is denatured in such a way that the individual chromosomes remain visually identifiable. The probe is labeled, either by incorporation of a ^{3}H-nucleotide or non-isotopically. The chromosomes are incubated with the labeled probe for many hours and, after hybridization, the excess unbound probe is washed away and the annealed probe detected. If the probe was radiolabeled, the slides are dipped into a photographic emulsion. The radioactive emission from the bound radiolabel interacts with the emulsion, depositing a silver grain. The grains will lie in the approximate area at which the probe bound to the chromosome. The accuracy of mapping by this technique is dependent on the size of the grains which, in the case of small chromosomes, can cover a significant portion of the total chromosome length.

If the probe is labeled non-isotopically then it is possible to determine the precise position of hybridization with much greater accuracy [4]. The signal is detected using a fluorescent molecule so that there is no spread of the signal such as occurs with autoradiography. This is called fluorescent *in situ* hybridization (FISH) (*Figure 2.8*). It is more sensitive than radiolabeling and can be used to give a fairly precise position of hybridization. Thus FISH is particularly useful for determining the respective locations of two markers on the same chromosome. The two probes are labeled with different immunologically labeled adducts and a single hybridization reaction is performed. The two probes are then detected using fluorochromes which emit at different wavelengths when excited by UV light, such as rhodamine (red fluorescence) and fluorescein (green fluorescence). This technique has been used to order probes along chromosomes when they are separated by no more than a few hundred kilobases of DNA. The FISH technique can also be applied to interphase nuclei, which allows accurate ordering of probes along a chromosome when they are separated by no more than 20–30 kb.

2.3.3 Genetic linkage mapping

This form of mapping is used to construct genetic linkage maps of individual chromosomes and also to localize genes involved in genetic diseases. Genetic linkage analysis relies on an ability to estimate the frequency of crossing over (recombination) that occurs between homologous chromosomes during meiosis. The closer two genetic markers are to each other then the lower the chance is that they will be separated during meiosis. The statistical probability that two markers are linked is measured as the logarithm of the odds (Lod). A Lod score of 3 is usually taken as the threshold of significance, while a Lod score below 2 suggests that the two loci are not linked. Lods between 2 and 3 are not considered significant.

2.4 Restriction enzyme mapping and its applications

The above three techniques allow a gene to be mapped to a specific chromosome, and linkage analysis and *in situ* hybridization allow its approximate position to be deduced relative to a known marker. However, more complex forms of chromosome analysis, which allow for the precise ordering of individual genes, are dependent on the ability to fragment the chromosomal DNA.

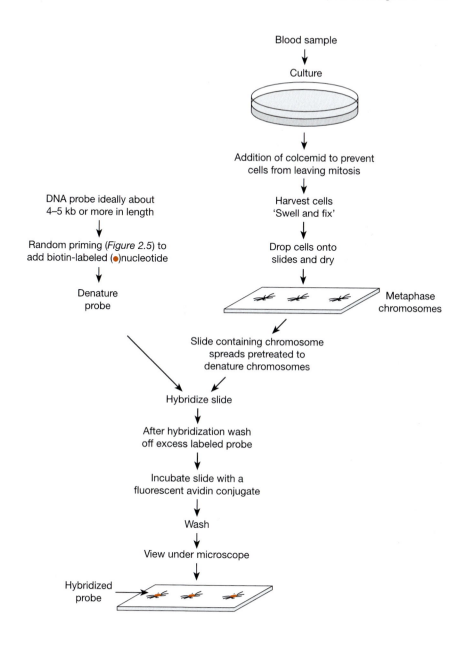

Figure 2.8. FISH is becoming the technique of choice in gene mapping experiments using *in situ* hybridization. It uses a non-radioactive labeling system which involves the incorporation of nucleotides with molecules of biotin attached. These have a strong affinity for avidin which in this example has been conjugated with a fluorescent molecule. When this is viewed under special filters the specific sites to which the DNA probe has bound can be seen on a chromosome.

2.4.1 Restriction enzymes

Restriction enzymes are prokaryotic proteins which cleave double-stranded DNA into discrete pieces, resolvable by gel electrophoresis. They cleave at, or very near to, specific recognition sequences within DNA. They are named after the species from which they derive (e.g. the enzyme *Hin*dIII derives from *Hemophilus influenzae*). Each species of bacterium contains a methylase which adds methyl groups to its DNA, protecting it from cleavage by its own restriction enzyme. Incoming foreign DNA lacks this modification and is destroyed by the restriction enzyme.

The restriction enzymes in common use recognize palindromic sequences within their DNA target. When read in a 5′ to 3′ direction, the sequence in each of the two complementary strands is identical (*Figure 2.9*). The recognition sequence is normally four, five or six nucleotides in length. The length and base composition of the recognition sequence determine its frequency of occurrence within any particular piece of DNA. Thus the four-nucleotide recognition sequence for the enzyme *Sau*3A, GATC, will occur once every 256 (4^4) nucleotides in DNA which contains equimolar amounts of the four bases. The six nucleotide recognition sequence for *Bam*HI, GGATCC, will occur once every 4096 nucleotides. Since they have the same four central residues, GATC, *Bam*HI will cut at a subset of *Sau*3A sites.

The frequency with which most restriction enzymes will cut a target DNA is predictable from the base composition of the target and the length and nucleotide sequence of the recognition site. However, certain cleavage sites are greatly under-represented in genomic DNA from mammalian cells. These are the sites which include the doublet CpG; for example, the recognition site for the enzyme *Sma*I which cuts at the sequence CCCGGG. The CpG dinucleotide is under-represented because it is a substrate for DNA methylation. Not all CpG dinucleotides within a particular tissue are modified and there is great variation between tissues in the extent of methylation at a particular CpG doublet. In general, a

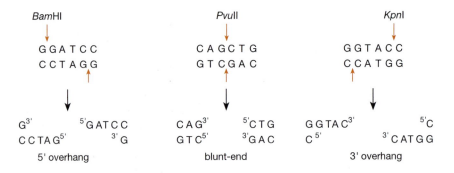

Figure 2.9. The three classes of cleavage sites in commonly used restriction enzymes.

gene is under-methylated in those tissues in which it is expressed. Thus in sperm cells the region of genomic DNA containing the globin genes is heavily methylated, while in erythroid cells it is much less highly modified [5].

When cytosine in a CpG dinucleotide is methylated, to form 5-methyl cytosine (*Figure 2.10*), there are two consequences. First, many enzymes will not cut a recognition site that is methylated in this way. Second, over evolutionary time, 5-methyl cytosine tends to be lost by spontaneous deamination. Unmethylated cytosine also deaminates, but the product is uracil, an unnatural base in DNA, which is recognized and repaired by the cell. 5-methyl cytosine, however, deaminates to the natural base thymine, which is therefore not repaired. Thus, mammalian genomic DNA contains far fewer CpG sequences than its overall base composition would predict, and many of those that are present are methylated. Because of this, enzymes whose recognition site includes one or more CpG sequences are known as rare-cutters. The enzyme *Not*I has an eight nucleotide recognition sequence GCGGCCGC with two CpG dinucleotides, and so it cleaves extremely infrequently in mammalian DNA. Such rare-cutter enzymes are particularly useful for creating large fragments containing long segments of genomic DNA for analysis by PFGE.

Some enzymes cut DNA at the same position in each strand, to generate flush or blunt termini (*Figure 2.9*). Others cut in an offset manner, to yield staggered termini. Since these are complementary, and hence mutually cohesive, they are often termed sticky ends.

Any piece of DNA of sufficient length will have recognition sites for one or more restriction enzymes and it is possible to determine the position of these sites. This is done by digesting the DNA with single enzymes, and with combinations of two enzymes (this is termed a double digest), and determining the sizes of the resultant fragments by gel electrophoresis (*Figure 2.11*). With a sufficient number of digests, it is possible to deduce the origin of each band on the gel; that is, which enzyme or combination

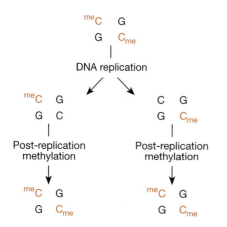

Figure 2.10. Inheritance of methylation at CpGs. This represents part of a double-stranded DNA molecule containing a CpG dinucleotide in which the C residue is methylated. When replicated, the two daughter molecules are initially methylated only in the parental strand (they are said to be hemi-methylated). A maintenance methylase then acts to modify the C residue in the newly synthesized strand.

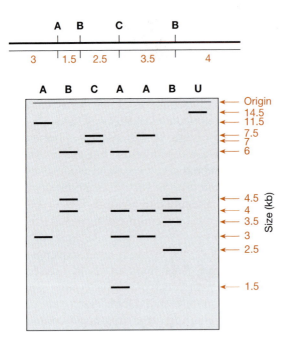

Figure 2.11. Restriction mapping of a piece of DNA. The DNA is shown in the upper part of the figure as a linear fragment containing restriction sites A, B and C, with two sites for enzyme B. By performing double digests with all combinations of the three enzymes, it is possible to deduce the relative positions of the cleavage sites and hence deduce the map. U is undigested DNA.

of enzymes generated it. This yields a restriction enzyme cleavage map, more commonly called a restriction map. Obtaining such a map is sometimes a first step in analyzing a piece of DNA of unknown structure.

2.4.2 Restriction enzyme mapping of genomic DNA

If total human genomic DNA is cut with a restriction enzyme such as *Eco*RI, which cuts on average about once every 4000 nucleotides, then approximately one million fragments are generated. When analyzed by agarose gel electrophoresis and stained with ethidium bromide, such a digest yields a smear extending throughout the gel. However, any individual *Eco*RI fragment will be of a specific size and it will therefore migrate as a single band. It is possible to detect such a band by performing a Southern transfer and hybridizing the filter with a probe complementary to some part of the fragment. Again, a restriction map of sites around the fragment can be built up by performing multiple digests of the total DNA. Here, however, in contrast to the situation described above for a single piece of DNA, it is only possible to identify the sites immediately flanking the probe (*Figure 2.12*).

2.4.3 Diagnosis of human genetic diseases using Southern analysis

Southern transfer has to a large extent been superceded by polymerase chain reaction (PCR)-based techniques in the diagnosis of inherited

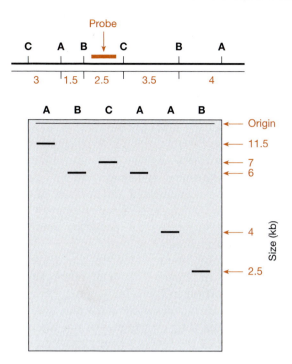

Figure 2.12. Restriction mapping by Southern transfer analysis of a piece of DNA forming part of a complex mixture of restriction fragments. In contrast to the situation described in *Figure 2.11*, where a discrete piece of DNA is analyzed on an ethidium bromide-stained gel, the DNA here is analyzed by Southern transfer using fragment B–C as a probe. This is the situation that arises when an entire mammalian genome is being analyzed to map a single gene from within it. It is only possible to detect the subset of fragments generated in the double digestion reactions which lie adjacent to the probe. For example, fragment C–A (at the extreme left-hand side of the map) will not be detected while the adjacent fragment A–C will be detected. Thus, for any one enzyme, the deduced map only gives information as to the nearest cleavage site upstream or downstream of a particular probe (i.e. an A site within the upstream C–A fragment would not be detected).

disease (PCR methods are described in Chapter 4). This is mainly due to the fact that PCR-based techniques are much more rapid than those based on Southern transfer. Southern analysis is, however, retained for analyses that are difficult or impossible to carry out by PCR. A particular category of human disease that poses problems for PCR analysis is the dynamic mutations. Dynamic mutations are trinucleotide repeats that, after an initial moderate expansion, become hypermutable, undergoing unpredictable further expansion. The upper limit of expansion in some of these diseases is quite modest (see Chapter 4 for more detail), < 100 repeats, and these mutations remain amenable to analysis by PCR. However, in other diseases the repeats can become much larger. It is these mutations that are difficult to analyse by PCR due to their sheer size, repetitiveness and in many cases GC content. Consequently they are analyzed by Southern transfer. One human disease where Southern transfer is still used is myotonic dystrophy (DM).

DM is a neuromuscular disease characterized in classically affected cases by muscle weakness, wasting and difficulty in relaxing muscles after contraction. DM is caused by a CTG trinucleotide expansion in the 3′ untranslated region of the *DMPK* gene [6]. Expansions above 50 repeats are considered to be associated with disease, the mildest disease being associated with small expansions of between 50 and 150 repeats. The number of repeats in classically affected and severely affected congenital cases can, however, rise to 2000 repeats. Since efficient PCR amplification through 2000 repeats is currently not possible, Southern transfer is retained as an alternative method of analysis.

One probe/enzyme combination used for analysis of large DM expansions is illustrated in *Figure 2.13*. The probe lies on an *Eco*RI fragment in which in addition to the CTG_n repeat there is an insertion/deletion polymorphism. In normal individuals the size of the genomic fragment is 9 or 10 kb depending on whether or not the polymorphism is present. Since expansions of the CTG_n repeat almost exclusively occur in association with the 10 kb allele, the presence of this polymorphism does not normally interfere with analysis. Consequently in individuals affected by DM there is a band that exceeds 10 kb in size (*Figure 2.14*).

Inversions that disrupt the internal structure of genes are also difficult to analyze by PCR, because they are unlikely to have their breakpoints within the coding sequence and they do not affect copy number. Although large inversions can often be detected by cytogenetic analysis, smaller ones can go unnoticed. Hence, Southern analysis can be very useful in detecting the presence of pathogenic inversions as novel restriction fragments. A common mutation within the Factor VIII gene is detected in this way.

Hemophilia A is a blood clotting disorder caused by the loss of activity of the coagulation factor VIII. It is caused by mutations within the Factor VIII gene. A common inversion mutation in the Factor VIII gene is found

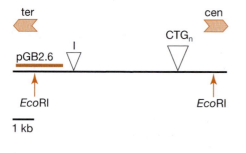

Figure 2.13. Restriction map of the genomic region surrounding the myotonic dystrophy CTG_n repeat. The two flanking *Eco*RI sites are shown as is the region of hybridization of the probe, pGB2.6, which detects a 9–10 kb fragment in normal individuals. The relative positions of the CTG_n repeat and the insertion/deletion polymorphism (I) are indicated.

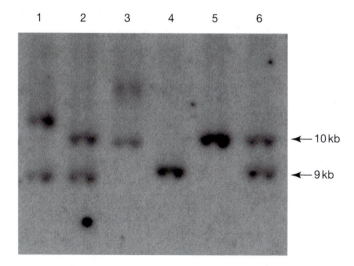

Figure 2.14. Southern analysis of large genomic expansions in myotonic dystrophy. Human genomic DNA was digested with *Eco*RI and probed with a ^{32}P labeled pGB2.6 probe (*Figure 2.13*). The positions of the 9 and 10 kb bands are shown. Lane 4 is a homozygote for the 9 kb deletion polymorphism, lane 5 is a homozygote for the 10 kb insertion polymorphism, and lane 6 is a heterozygote for the insertion/deletion polymorphism. Lanes 1 and 3 are both carriers of large expansions as shown by fragments greater than 10 kb in size. Lane 2 is a carrier with a small expansion (approximately 80 repeats).

in over 40% of severely affected cases [7]. It involves the recombination of homologous sequences that are present within an intron of the Factor VIII gene and that are also present within the promoter of the Factor VIII gene. Recombination between these sequences leads to an inversion that disrupts the organization of the Factor VIII gene (*Figure 2.15*). This causes Hemophilia A in male individuals and the inversion can be detected by Southern transfer analysis (*Figure 2.16*).

2.4.4 Variable number tandem repeats in paternity testing and forensic medicine

Variable number tandem repeats (VNTRs) are short tandemly repeated DNA sequences, where the number of repeat units at a particular chromosomal location varies from person to person. The size of different repeated units varies from a single nucleotide up to tens of nucleotides. Repeats of 1–4 nucleotides are called microsatellites and their special features are discussed in Chapter 4. The major advantage of VNTRs over restriction fragment length polymorphisms (RFLPs) is that VNTR markers show multi-allelic variation. Rather than just two alleles, giving rise to only three possible combinations of alleles in a lineage, there is great variation in the size of each VNTR within the genome of different

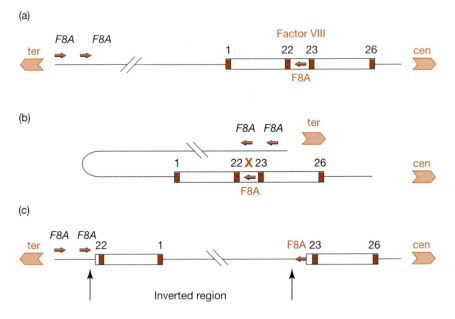

Figure 2.15. Origin of the common Factor VIII inversion mutation. (a) The normal structure of the Factor VIII gene. The Factor VIII gene consists of 26 exons; the relative positions of exons 1, 22, 23 and 26 are illustrated. Within intron 22 there lies a gene transcribed on the opposite strand to Factor VIII, F8A. Upstream of exon 1 of Factor VIII, the gene towards the telomere of the X-chromosome, lie two highly homologous copies of F8A labeled on the diagram as F8A. (b) Occasionally pairing can occur between one of the distal copies of F8A and the intron 22 copy F8A by looping back of the chromatid. (c) Chromatid breakage followed by misrepair can give rise to an inversion of the whole region, leading to disruption of the Factor VIII gene. Two inversions are possible depending on which F8A copy is involved.

individuals. This high level of individual variation is the basis for their use in paternity and identity testing in forensic medicine. When a combination of four or five VNTR markers is used they can confirm identity with a high level of certainty.

The cause of the high level of allelic variation between VNTRs has been controversial. One theory ascribes it to unequal crossing over. At meiosis the pairs of chromosomes come together and recombination leads to an exchange of genetic material between homologs. If, during this crossing over, the chromosomes are not aligned perfectly, there will be a non-reciprocal exchange of material. This is termed unequal crossing over (*Figure 2.17*). The presence of tandem repeats in the genome could lead to misalignment and unequal crossing over, which would perpetuate and extend the degree of allelic variation between VNTRs. Although this explanation sounds plausible, evidence for recombination has been hard to find. It now seems more likely that variants arise during DNA replication, by slippage or 'stuttering' of the polymerase.

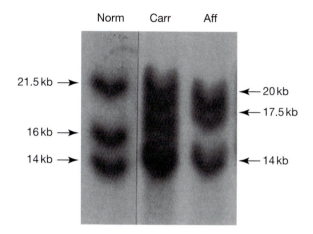

Figure 2.16. Southern analysis of the common Factor VIII inversion. Two inversion types are possible depending on which of the two upstream F8A copies was involved in the mispairing event. Southern analysis can detect the presence of both types of inversion. Hybridization of the probe p482.6a to *Bcl*I digested genomic DNA gives rise to fragments of 21.5, 16 and 14 kb in normal alleles, 20, 17.5 and 14 kb in distal F8A inversions and 20, 16 and 15.5 kb in proximal inversions (*Figure 2.15*) permitting easy diagnosis. *Bcl*I digest of genomic DNA hybridized with the probe p482.6a. Three individual samples are illustrated: Norm – the pattern derived from a normal individual; Aff – the pattern derived from a male individual with an inversion involving the distal most F8A copy and thus affected by hemophilia A; and Carr – the pattern derived from a female carrier of a distal F8A inversion.

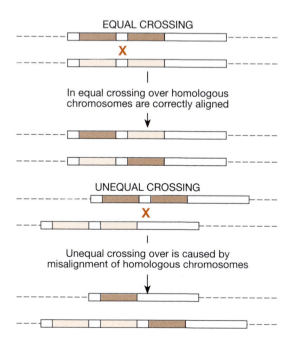

Figure 2.17. Unequal crossing over: a possible mechanism for generating VNTRs.

If a Southern blot analysis is undertaken using a probe that detects a restriction fragment containing a VNTR, then different individuals in the population will display two different sized restriction fragments, one from the maternally inherited chromosome and the other from the paternally derived chromosome. Changes in the repeat number of some VNTRs occur much more frequently than other DNA sequence changes, such as point mutations or deletions of non-repetitive sequences. However, in general, highly variable VNTRs are sufficiently stable to be useful genetic markers. Thus, provided that a VNTR can be found within a reasonably close distance of a gene of interest, VNTRs can be used to diagnose genetic diseases. The first VNTR to be described [8] was assigned to the telomeric region of the long arm of chromosome 14. Initially, VNTRs were thought to be distributed randomly throughout the genome but, as more have been accurately mapped, it has become evident that they tend to be clustered towards chromosomal telomeres.

For genetically unrelated individuals, the precise sizes of the two bands observed will almost always be different. Even when individuals are closely related genetically, they can be distinguished by the use of several different probes. This is not true of conventional RFLPs, because there are only a restricted number of variants within any population. Thus VNTRs provide the first true 'molecular fingerprint' which, with the exception of monozygotic twins, is unique for every member of the human race. There are two very important applications of this fact – paternity testing and forensic medicine.

If pedigree analysis is performed on an individual using many different VNTR polymorphisms, then every band (allele) seen in that person's DNA should be present in one or other of the parents. Each parent will of course also display some length variants that are not present in their child's genome, because the child inherits only half of each parent's genes, but if the child has several bands not seen in the DNA of the supposed parents, then it is highly unlikely that the supposed parentage is correct. Usually the explanation is non-paternity, but alternatives are unacknow-ledged adoption or babies mistakenly switched at birth. Exclusion of alleged paternity is thus straightforward: if the child shares few or no bands with the alleged father, and has bands not present in him or the mother, then the alleged father is not the true father. Thus VNTRs provide a perfect method of paternity testing and there are now commercial companies that provide such services.

With suitable VNTRs the spectrum of bands is so variable that the chance of any other individual in the population coincidentally sharing a high fraction of variants can be extremely low. This allows positive assignment of paternity with very high probability (but never 100%), something which is not possible with any other class of genetic markers. Because the spectrum of VNTRs is unique to an individual, they also provide enormously useful tools for forensic identification. Provided only that sufficient tissue is left at the scene of a crime, then VNTR analysis can be used to distinguish between suspects. For Southern analysis sufficient tissue must be available to extract microgram quantities of DNA. This would be a serious limitation if blood or semen were to be used in the analysis. Fortunately, however, there exists a very much more sensitive analysis method called PCR, which can be applied to vanishingly small tissue samples. Description of this method requires a knowledge of fine structure mapping techniques, and this is the subject of the next chapter.

References

1. **Southern, E.M.** (1975) Detection of specific sequences among DNA fragments separated by gel electrophoresis. *J. Mol. Biol.* **98**: 503.
2. **Alwine, J.C., Kemp, D.J. and Stark, G.R.** (1977) Method for detection of specific RNAs in agarose gels by transfer to diazobenzyloxymethyl-paper and hybridization with DNA probes. *Proc. Natl Acad. Sci. USA* **74**: 5350.
3. **Feinberg, A.P. and Vogelstein, B.** (1984) A technique for radiolabeling DNA restriction endonuclease fragments to high specific activity. *Anal. Biochem.* **137**: 266.
4. **Trask, B.J., Massa, H., Kenwrick, S. and Gitschier, J.** (1991) Mapping of human chromosome Xq28 by two-color fluorescence in situ hybridization of DNA sequences to interphase cell nuclei. *Am. J. Hum. Genet.* **48**: 1.
5. **Barker, D., Schafer, M. and White, R.** (1984) Restriction sites containing CpG show a higher frequency of polymorphism in human DNA. *Cell* **36**: 131.
6. **Brook, D.J., McCurrach, M.E., Harley, H.G.** *et al.* (1992) Molecular basis of myotonic dystrophy: expansion of a trinucleotide (CTG) repeat at the 3' end of a transcript encoding a protein kinase family member. *Cell* **68**: 799–808.
7. **Van de Water, N.S., Williams, R., Nelson, J.** (1995) Factor VIII gene inversions in severe hemophilia A patients. *Pathology* **27**: 83–85.
8. **Wyman, A.R. and White, R. and Browett, P.J.** (1980) A highly polymorphic locus in human DNA. *Proc. Natl Acad. Sci. USA* **77**: 6754.

Further reading

Anand, R.E. (1992) *Techniques for the Analysis of Complex Genomes.* Academic Press, San Diego, CA.
Antonarakis, S.E. (1989) Diagnosis of genetic disorders at the DNA level. *N. Engl. J. Med.* **320**: 153.
Botstein, D., White R.L., Scolnick, M. and Davis, R.W. (1980) Construction of genetic linkage map in man using restriction fragment length polymorphisms. *Am. J. Hum. Genet.* **32**: 314.
Davies, K.E. and Tilghman, S.M. (1991) *Genome Analysis*, Vol. 1, *Genetic and Physical Mapping.* Cold Spring Harbor Press, New York.
Elles, R. (1996) *Molecular Diagnosis of Genetic Diseases.* Humana Press, Totowa, NJ.
Glover D.M. (1990) *DNA Cloning: a Practical Approach.* IRL Press, Oxford.
Marx, J. (1989) *A Revolution in Biotechnology* Cambridge University Press, Cambridge.
Wetherall, D.J. (1991) *The New Genetics and Clinical Practice*, 3rd Edn. Oxford University Press, Oxford.

Winnacker, E.-L. (1989) *From Genes to Clones, Introduction to Gene Technology.* VCH Verlag, Weinheim.

Rubinsztein, D.C. and Hayden, M.R. (1998) *Analysis of Triplet Repeat Disorders.* BIOS Scientific Publishers, Oxford.

Chapter 3

Gene analysis techniques II: determining the structure of genes

3.1 DNA sequence analysis

Restriction enzyme mapping and Southern blotting help establish the gross organization of a piece of DNA, and allow the approximate position of sequences of interest to be determined. Normally, the next step in the analysis is to determine the nucleotide sequence. These techniques are constantly evolving and they will be described in a semi-historical manner, because this will illustrate the basic principles and also because some of the original techniques still have important uses. Techniques for determining DNA sequence initially became possible because of the remarkable resolving power of polyacrylamide gels run in the presence of urea as a denaturant. If two single-stranded DNA molecules, identical in sequence over almost all of their length but differing by the presence of just one additional base on one of the two molecules, are subjected to electrophoresis in a denaturing polyacrylamide gel, they will be separated. Two DNA sequencing technologies were devised, based upon this principle: the chemical degradation procedure and the chain termination procedure [1,2].

3.1.1 Sequencing by chemical degradation

In the chemical degradation ('Maxam and Gilbert') procedure [1], a DNA restriction fragment is isotopically labeled at either its 5′ or its 3′ terminus and each of the two strands is separated and isolated (*Figure 3.1*). The DNA is then partially modified with chemical reagents specific for the different bases and cleaved at the modified nucleotides. This generates a set of molecules differing in length but with the same isotopically labeled terminus. These fragments are subjected to denaturing gel electrophoresis in parallel slots of a high-resolution polyacrylamide gel,

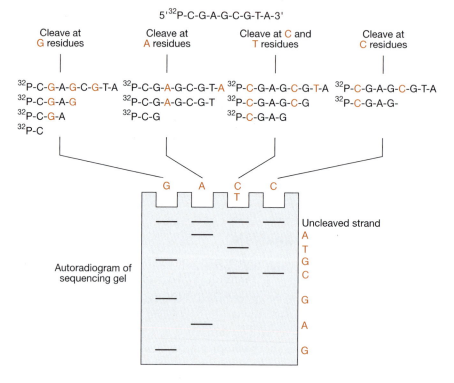

Figure 3.1. DNA sequence analysis by the chemical degradation (Maxam and Gilbert) technique.

and the sequence of the DNA can be deduced from the resulting autoradiogram. Because it is a relatively laborious procedure that is not readily automated, this technique is now only used in certain, specialized gene analysis techniques.

3.1.2 Sequencing by chain termination

The chemical degradation procedure has been almost totally replaced as a sequence determination tool by the much more convenient chain termination ('Sanger') method [2]. In this procedure DNA is synthesized rather than degraded, but the fundamental principle is similar to the chemical method. DNA polymerases function by extending a pre-existent DNA strand, the primer, which is annealed to the template strand in the configuration shown in *Figure 3.2*. In the chain termination method, a short synthetic oligonucleotide is used as the primer and it provides a fixed 5′ terminus for the growing chain. The reaction is performed in the presence of chain terminators. These are nucleotide derivatives which

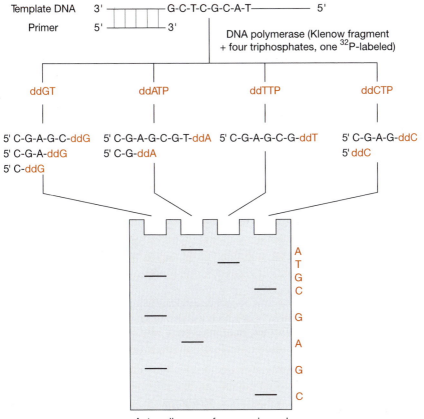

Figure 3.2. DNA sequence analysis by the chain termination (Sanger) technique.

contain a blocking group at their 3′ hydroxyl terminus. Once incorporated into the growing chain, they prevent further extension. A small amount of chain terminator, for example dideoxy GTP (*Figure 3.3*), is mixed with unmodified GTP and the other three nucleoside triphosphates. In this instance, a series of reaction products is produced which terminates at a subset of G residues in the newly synthesized strand (*Figure 3.2*). In the original, radioisotopic detection procedure [2], four parallel reactions are run, each including a dideoxy chain terminator for one of the four nucleotides, and the products analyzed in parallel slots of a denaturing polyacrylamide gel. The growing chain is rendered detectable by including a radiolabeled nucleoside triphosphate in the synthesis reaction. As with the chemical cleavage procedure, the nucleotide sequence of the template strand can be deduced by comparison of the pattern of autoradiographic bands in the four lanes of the gel (*Figure 3.4*).

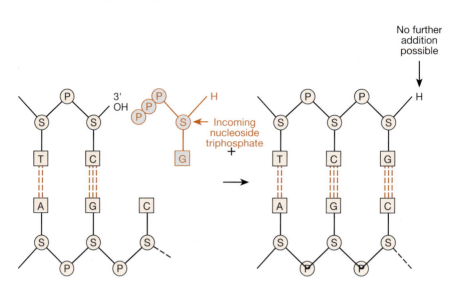

Figure 3.3. The chemical basis of sequence analysis using chain terminators. This is a representation of a reaction in which a dideoxy G residue incorporated into the growing chain blocks further incorporation.

3.1.3 Automated DNA sequence analysis

Using the chain termination method, DNA sequence can be determined very rapidly and there are now analysis methods and machines based upon this approach which speed the process even further. These techniques have almost entirely replaced the isotopic labeling method, because they are much faster, easily automated and there is no radiation hazard. In a conventional sequencing reaction four separate single-strand extension reactions are performed, each containing a different chain-terminating dideoxynucleoside triphosphate. The radiolabeled reaction products are separated on four separate tracks of a polyacrylamide gel. In machine-based sequencing a single reaction is run, with each of the four dideoxynucleoside triphosphates containing a different fluorescent label. When excited by an argon laser, the fluorescent labels each emit light at a different wavelength. The same chain-termination reactions are carried out as with the isotopic method, but it is no longer necessary to run the products in separate lanes of a gel. The reaction products are mixed and run in a single lane in a gel and, as the different length fragments pass the laser, each fluorescent label gives a unique signal, which is stored in a computer (*Figure 3.5*).

Alternatively the reaction products can be resolved on a chromato-graphic column containing linear, unpolymerized acrylamide or on an

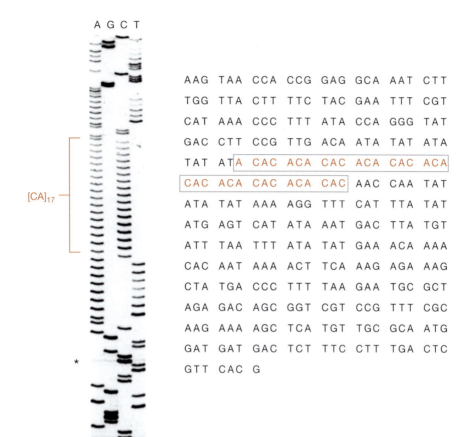

AAG TAA CCA CCG GAG GCA AAT CTT

TGG TTA CTT TTC TAC GAA TTT CGT

CAT AAA CCC TTT ATA CCA GGG TAT

GAC CTT CCG TTG ACA ATA TAT ATA

TAT AT A CAC ACA CAC ACA CAC ACA

CAC ACA CAC ACA CAC AAC CAA TAT

ATA TAT AAA AGG TTT CAT TTA TAT

ATG AGT CAT ATA AAT GAC TTA TGT

ATT TAA TTT ATA TAT GAA ACA AAA

CAC AAT AAA ACT TCA AAG AGA AAG

CTA TGA CCC TTT TAA GAA TGC GCT

AGA GAC AGC GGT CGT CCG TTT CGC

AAG AAA AGC TCA TGT TGC GCA ATG

GAT GAT GAC TCT TTC CTT TGA CTC

GTT CAC G

Figure 3.4. Typical DNA sequence data obtained using the chain termination method. This shows the sequence of a region of a gene that contains a repeat of the sequence CA. Most of the sequence is completely unambiguous but the positions of some artifactual bands are indicated by *. Here there are bands in all four lanes, although in each case one band is much more intense than the other three. This band would usually be provisionally interpreted to be the 'correct' nucleotide at this point in the chain, e.g. the top-most indicated artifact would be interpreted to be a C residue. It is normally essential to determine the sequence of both strands of a DNA molecule, so that such provisional assignments would hopefully be confirmed when the other strand was sequenced.

array of such columns. In the case of both gel-based and column-based machines the DNA sequence is visualized as a series of peaks that represent the fluorescence signal for the four dyes (*Figure 3.6*).

A computer program then analyses the peaks and produces a suggested sequence, but ambiguities in the sequence need to be resolved by manually inspecting the raw data.

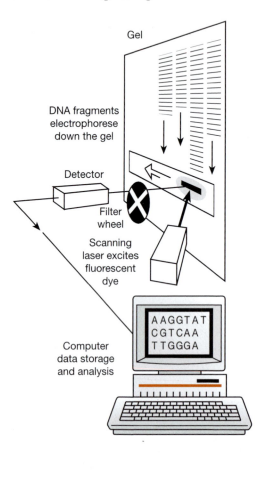

Gel

DNA fragments
electrophorese
down the gel

Detector

Filter
wheel

Scanning
laser excites
fluorescent
dye

Computer
data storage
and analysis

AAGGTAT
CGTCAA
TTGGGA

Figure 3.5. Automated DNA sequence analysis. The chain terminator method of DNA sequencing has been modified to allow DNA sequencing to be automated. Fluorescent dyes are attached to each of the dideoxynucleotides, replacing the need for radioactive nucleotides. After the sequencing reactions have been completed the products of all four reactions are pooled and electrophoresed in one lane of a polyacrylamide sequencing gel. As the reaction products pass through the gel the specific DNA bases are detected by a scanning laser that excites each fluorescent dye at different wavelengths. This information is captured on a detector and passed to a computer for data storage and analysis. Alternatively, the reaction products can be separated electrophoretically on a column.

3.1.4 Preparation of templates for DNA sequencing

DNA sequence analysis using chain terminators is optimally performed using a single-stranded template. This allows the primer unhindered access to the template. A suitable template for single-strand DNA sequence analysis can be created by inserting the double-stranded template DNA into the genome of a bacterial virus which naturally packages only one of the two strands into viral particles (see Sections 5.2.1 and 5.3.1). Insertion of the DNA into such a vector does, however, add another step to the sequencing procedure, and this is not always convenient. If a double-stranded DNA template is denatured by treatment with alkali, and an excess of primer is added before re-annealing occurs, the primer will compete with the complementary strand for annealing to the template. This arrangement allows double-strand DNA sequence analysis to be performed, and this is now very often the method of choice.

The major disadvantage of double-stranded relative to single-stranded DNA sequence analysis is that the presence of the competing comple-

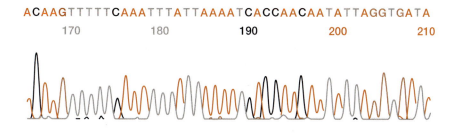

ACAAGTTTTTCAAATTTATTAAAATCACCAACAATATTAGGTGATA

170   180   190   200   210

Figure 3.6. Typical output from a DNA sequence analysis using fluorescent dyes.

mentary strand leads to a higher level of sequencing artifacts. These are positions on the gel where more than one band is present (e.g. see *Figure 3.4*), so that it may be impossible to decide which is the correct residue at that position. Artifacts of this kind are not restricted to double-strand DNA sequence analysis; some level of artifact is a universal problem in sequence analysis. The resultant ambiguities can often be resolved only by re-sequencing the template using another primer, or by using the other strand (i.e. the strand which is complementary to that used in the original analysis) as template. Even using a single-stranded DNA as template, it is not usually possible, on either a gel or column, to cleanly resolve reaction products when the molecules are much larger than 600–700 nucleotides. This sets the upper limit to the length of sequence which can be determined using a particular primer–template complex. One way to extend from a known into an unknown region of sequence is to synthesize an oligonucleotide primer which is complementary to a tract near the end of the region whose sequence has already been established. In this way it is possible to 'walk' along a sequence, at each step synthesizing a new primer which is used to extend the region of known sequence (*Figure 3.7*).

Alternatively, it is possible to delete sequences enzymatically from the template at a position downstream of the annealing site of the primer, to generate a deletion series (*Figure 3.8*).

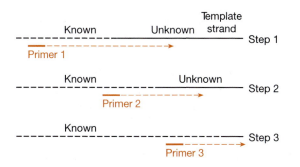

Figure 3.7. DNA sequence analysis by 'primer walking'.

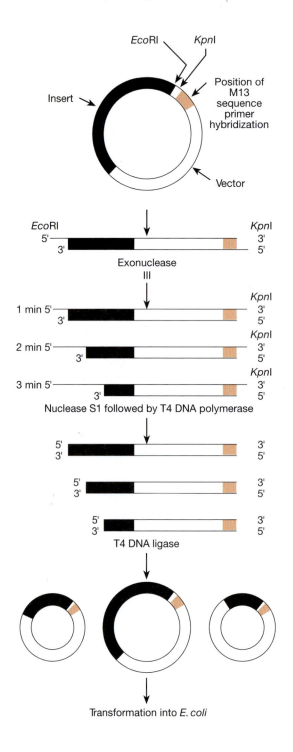

Figure 3.8. The use of nested deletions for DNA sequence analysis.

A method that has become very important in the sequence analysis of whole genomes dates back to the very early days of sequence analysis and is known as shotgun sequencing (*Figure 3.9*). The DNA to be sequenced is randomly fragmented and the pieces are cloned into a plasmid or M13 vector (see Chapter 5). The fragments are then sequenced, from one end or sometimes from both ends, by priming from within the vector DNA adjacent to the site of insertion. Sequencing is continued to the point where the total amount of sequence determined is 5–10 times greater than the length of the starting DNA template. A computer is then used to establish how the different sequence reads overlap one another and this information can be used to predict the sequence of the starting DNA molecule. The large number of shotgun sequence reads performed, which have different start and end points and that are read in different directions, ensure the accuracy of the established sequence.

By determining the sequence of both strands of a DNA molecule, and confirming that the two deduced sequences are perfectly complementary, it is possible to detect and eliminate artifacts generated in the reactions or during electrophoresis. When this check is performed, DNA sequence can be established with a very high degree of accuracy. Machine-based DNA sequencing has radically changed the speed and scope of sequencing projects. At the time of writing, whole genome sequences are known for many bacteria, for the yeast *Saccharomyces cerevisiae* and for *Caenorhabditis elegans*. Also, there is a 'working draft' of the human genome sequence, with many more genomic sequences well on the way to being established.

3.2 Transcript mapping techniques

Analyzing the nucleotide sequence of a piece of DNA is a necessary preliminary to understanding its precise organization and function, but is

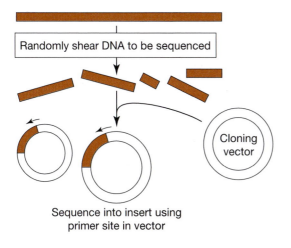

Figure 3.9. DNA sequence analysis by the shotgun method.

not normally sufficient for these purposes. Because eukaryotic gene transcripts are subjected to a series of post-transcriptional processing events, the structure of a gene can be understood only by correlating the organization of the genomic DNA with the organization of its mRNA transcript. This has required the establishment of cloning and mapping techniques which can, when used in conjunction, give a complete description of the structure of a gene.

It is possible to isolate, by gene cloning, the double-stranded DNA copies of individual mRNA sequences (cDNAs) (see Chapter 6). Such 'cDNA clones' will normally contain a DNA copy of the poly(A) sequence at the 3′ terminus. Hence, by determining which strand of the cDNA clone contains a poly(A) tract at one end, it is possible to identify the coding strand of the cDNA clone. Provided a sufficient length of sequence has been established to make their presence statistically likely (in practice 300–500 nucleotides), termination codons will occur in two of the three potential coding frames, ruling them out as translational open reading frames (ORFs). Sometimes this level of analysis, that is, identification of the reading frame and deduction of the encoded protein sequence, will be all that is required. However, it is often desirable to determine the structure of the gene itself. It will have been isolated as part of a genomic clone (Chapter 5). This will contain the gene and a variable amount of flanking DNA depending upon the method of cloning used.

By correlating the sequence of the gene with the sequence of the mRNA deduced from the cDNA clone, it is possible to identify the site of poly(A) addition and the positions of introns. However, technical limitations in the cDNA cloning procedure ensure that the cDNA clone will not normally contain sequences derived from the extreme 5′ terminus of the mRNA. There are two methods to determine the position of the 5′ end: primer extension and nuclease protection.

3.2.1 Primer extension

In primer extension, a short oligonucleotide complementary to a sequence tract near to the 5′ end of the mRNA is annealed to the mRNA. This is then used to prime synthesis with the enzyme reverse transcriptase (*Figure 3.10*). This enzyme produces a cDNA which extends to the 5′ terminus of the mRNA. By correlating the length of the cDNA primer extension product with the sequence of the genomic DNA, it is possible to deduce the start site of transcription. A related approach to mRNA 5′ end mapping is the rapid amplification of cDNA ends (RACE) technique. In this method, primer extension is combined with a PCR step (see Section 3.4).

3.2.2 Nuclease protection

An alternative technique for transcript mapping is called nuclease protection (*Figure 3.11*). An isotopically labeled, single-stranded RNA

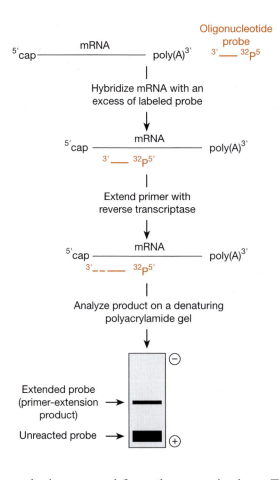

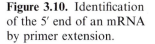

Figure 3.10. Identification of the 5′ end of an mRNA by primer extension.

probe is generated from the genomic clone. The probe is designed to be complementary to the mRNA over a region encompassing part or all of the first exon, and extending upstream of the region where the cap site is expected to be located. The cap site will often have already been putatively localized using primer extension. The probe is annealed to the mRNA and digested with a mixture of nucleases that destroy single-stranded RNA but leave double-stranded RNA molecules intact. (Note, in early studies end-labeled DNA probes were used in conjunction with an enzyme called nuclease S1; hence the name S1 mapping is sometimes used, even though RNA probes are now almost invariably employed.) During digestion, the 3′ proximal region of the probe, which has no complement in the mRNA molecule, and which therefore remains single-stranded, is degraded. A molecule is generated which is the complement of the region between the cap site and the position at which the 5′ terminus of the probe annealed to the mRNA. Again, by determining the length of this molecule it is possible to infer the position of the cap site.

The above two techniques are often used in parallel to determine the cap site. They are both subject to artifacts, but of different kinds.

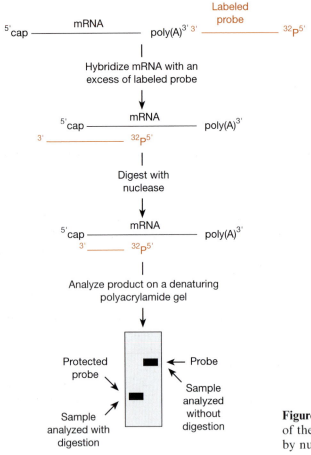

Figure 3.11. Identification of the 5′ end of an mRNA by nuclease protection.

Therefore, if both methods agree, it greatly increases confidence in the assignment of the cap site. If the hybridization is performed with an excess of probe, the amount of primer extension or nuclease-resistant product is proportional to the amount of complementary mRNA. Both techniques can therefore be used as an alternative to Northern transfer to quantitate mRNA levels.

3.3 Searching for genes using computers

Although the series of analyses described above are necessary, ultimately, to establish the organization and mode of expression of a particular gene, genes can be sought by less laborious means using computer programs. These search for regions of a DNA sequence which look like coding regions flanked by the signals that direct gene transcription and processing of their RNA products (*Figure 3.12*).

This procedure is of greatest value when large regions of genomic DNA for which cDNA clones are not available are being analyzed. It is thus an

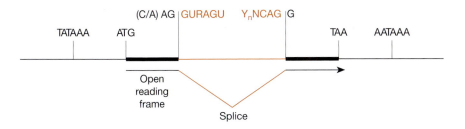

Figure 3.12. Searching for genes using computers. Structure of a gene shown as consensus sequences.

essential part of the Human Genome Project. One of the principles, that of searching for long ORFs, has already been described. Although the sequence (the Kozak sequence) that precedes the ATG initiation codon is only weakly conserved between different genes it can provide a helpful guide to the start site of translation. The start site of transcription often lies downstream of a TATA box, and it is also possible to search for the consensus binding sites of other transcription factors. Binding sites for the transcription factor Sp1 (Chapter 1) are particularly common, both immediately upstream of 'housekeeping' genes and within their 5' proximal coding regions. A consensus Sp1 site contains a CpG dinucleotide, a dinucleotide which is greatly under-represented in other parts of the mammalian genome. Hence the 5' end of genes which have such a structure are sometimes called CpG islands [3]. These regions can be identified, even in the absence of DNA sequence information, because they are susceptible to cleavage by rare-cutting restriction enzymes such as *Not*I (Chapter 2). Their presence at the 5' end of a long ORF, or series of ORFs, is a strong indicator that a gene is present.

RNA processing signals can also be used to help establish the likely gene organization. The sequence that signals polyadenylation is well conserved, and splice junctions have a consensus sequence comprising an almost perfectly conserved core (5' GU– – – –AG 3'), with moderately well conserved flanking sequences (*Figure 3.12*). Identification of genes by analysis of DNA sequences is an evolving science. The structures of new genes are constantly adding to the rules, and the sophistication of the search programs is increasing. However, at present it is used largely as a guide to further action, that is, to the application of the precise mapping techniques described above. Such deduced gene structures can, however, be used to determine whether a similar gene has ever been observed before. This is done by computer searches of protein or DNA databases. These databanks contain many thousands of gene sequences. The sizes of the banks are such that there is a reasonable likelihood of finding a homolog to any gene sequence used in the search. Even if the whole of the gene has no homolog in the database, very often a part of the gene will be homologous to another gene. This happens because genes are generally built up of modules, for example, an actin-binding domain linked to a calcium-binding domain, or a DNA binding domain linked to a potential transcriptional activation domain. Identifying a module within a gene can be an invaluable clue to its likely function.

3.4 The polymerase chain reaction

PCR has become one of the most valuable techniques in molecular biology and underlies many of the most exciting advances made in recent years [4]. Relying only upon a knowledge of the nucleic acid sequence of a gene and its functional organization, it allows the synthesis of microgram amounts of specific nucleic acid sequence from any part of the genome. It can, for example, be used as a powerful diagnostic test for the presence of mutations within the human genome, or to introduce specific mutations into a cloned gene.

In PCR, two oligonucleotide primers are synthesized which derive from opposite strands of the template DNA to be amplified and which have their 3′ termini facing each other (*Figure 3.13*). The target DNA is denatured in the presence of a large excess of the two primers and then returned to a temperature which will allow the primers to anneal to the DNA. A heat-stable DNA polymerase (most frequently that isolated from the thermophilic bacterium *Thermus aquaticus* and therefore called *Taq* polymerase) and all four nucleoside triphosphates are included in the reaction and the sample is placed at a temperature that is optimal for elongation by the enzyme. This produces two copies of the sequence lying between the primers. If the reaction has occurred for a sufficient time, the sequences will extend beyond the position at which the oppositely oriented primer annealed.

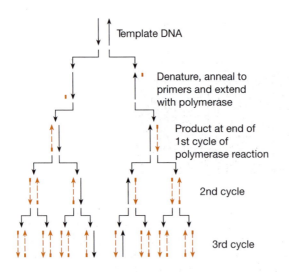

Figure 3.13. The polymerase chain reaction.

The above three steps of melting, annealing and extension are repeated. This produces another four copies of the sequence bracketed by the oligonucleotide primers. Again, two of these will have indeterminate 3′ termini but two will now have termini dictated by the 5′ terminus of the other primer. In the subsequent cycles of reactions there will be further amplification such that, within a few cycles, the predominant product is that defined by the 5′ termini of the starting primers. Originally, PCR was performed manually, by laboriously transferring the test tubes between water baths at the required temperatures. Now, however, commercially available machines are used in which the tubes sit within a metal heating block that is programed to cycle rapidly between the required temperatures.

In principle, n cycles of PCR will amplify the target 2^- fold. *Taq* polymerase is very thermostable, and 50 or more cycles of amplification can easily be performed if needed. Hence, as little as a single molecule of target can generate a detectable amount of PCR product. Initially, the size of the DNA to be amplified was a major factor limiting the yield of PCR reaction product but now there are commercially available enzymes that will amplify targets as large as 40 kb. PCR is therefore an analytical technique of immense power [5,6]. In fact, one of the major problems to be overcome when using PCR at this level of detection is oversensitivity. False positives are all too readily generated by trace amounts of the target DNA in the laboratory environment. Again, because the technique is so sensitive, there is also a risk of generating false positives from the target added to the tube. This occurs when one of the primers cross-hybridizes to sequences within the target DNA that bear some degree of homology to the intended target sequence. This can, in the worst case, lead to a whole welter of bands on the analytical gel, only one of which is the authentic reaction product. This problem can usually be resolved by altering the PCR conditions to minimize cross-hybridization. Since the ionic strength is one of the factors determining T_m, this can be conveniently done by decreasing the magnesium concentration and monitoring the reaction products (*Figure 3.14*).

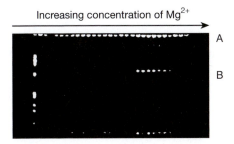

Figure 3.14. PCR reaction showing the effect of increasing concentrations of magnesium in the reaction buffer. A and B are two different genes amplified together and showing considerable variation in the amount of product produced depending on magnesium concentration. C are oligonucleotide primers running at the front of the gel. The PCR products were separated on an acrylamide gel and visualized under UV light after staining with ethidium bromide.

Determining the sequence of PCR products, isolated as individual chains by molecular cloning, has revealed that there is an error rate in their synthesis. This can be as high as one misincorporated base per thousand synthesized [7]. This is not a problem when PCR is being used analytically, since the size of the product and/or its ability to hybridize to a complementary DNA probe will not be affected. It can, however, be a problem when the aim is to clone the PCR product for some specific use. The only absolutely safe procedure is to determine the entire DNA sequence of several clones containing the PCR product to check for errors. One method of sequencing a PCR product is shown in *Figure 3.15*.

The PCR reaction has found use in many genetic engineering techniques and some of these will be presented in later chapters. One

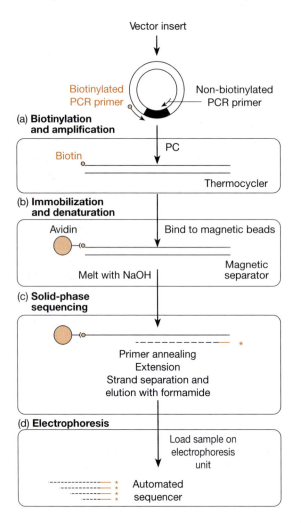

Figure 3.15. PCR sequence analysis using biotinylated oligonucleotides. This helps produce PCR products which are suitable as templates for sequencing. One of the two PCR primers is modified by incorporating a biotinylated deoxynucleotide as the final base during synthesis of the primer. After PCR with the modified primers, the product is isolated by adding streptavidin-coated magnetic beads to the reaction mix. Biotin has an extremely strong affinity for streptavidin, and the biotinylated product will bind to the magnetic beads. The beads are then purified away from the reaction using a magnet. This is a double-stranded product; single-stranded template for sequencing is obtained by mild treatment with alkali. The strand bound to the beads is washed and used as template for chain termination sequencing.

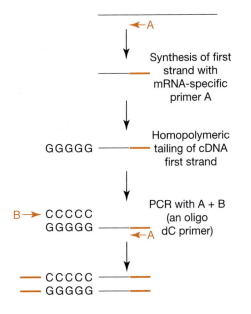

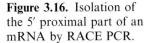

Figure 3.16. Isolation of the 5′ proximal part of an mRNA by RACE PCR.

PCR-based technique, that is used in mRNA mapping (see Section 3.2.1), is the RACE PCR method for determining the start site of transcription. An oligonucleotide primer (A) is designed that is complementary to a known region of sequence near the 5′ end of the mRNA. After synthesis of a first (cDNA) strand, a homopolymeric tract of G residues is added to the cDNA enzymatically (*Figure 3.16*). Then a second primer (B), incorporating a stretch of Cs at its 5′ end, is used to PCR amplify the region in between primers A and B. The PCR product is then either cloned into a plasmid vector (Section 5.2) or its sequence is determined directly (*Figure 3.15*). By comparing the sequence of the PCR product with the relevant genomic sequence, it is possible to determine the position of the transcriptional start site. [Note, RACE PCR is actually more frequently used to obtain a cDNA segment representing the 5′ proximal part of an mRNA (see Section 6.1.2) in a situation where the relevant genomic sequence is not already known.]

References

1. **Maxam, A.M. and Gilbert, W.** (1977) A new method for sequencing DNA. *Proc. Natl Acad. Sci. USA* **74**: 560–564.
2. **Sanger, F., Coulson, A.R., Barrell, B.G., Smith, A.J.H. and Roe, B.A.** (1980) Cloning in single-stranded bacteriophage as an aid to rapid DNA sequencing. *J. Mol. Biol.* **143**: 161–178.
3. **Bird, A.P.** (1986) CpG-rich islands and the function of DNA methylation. *Nature* **321**: 209–213.
4. **Saiki, R.K., Scharf, S.J., Faloona, F.** *et al.* (1985) Enzymatic amplification of beta-globin genomic sequences and restriction site analysis for diagnosis of sickle cell anemia. *Science* **230**: 1350–1354.

5. **Cheng, S., Fockler, C., Barnes, W.M. and Higuchi, R.** (1994) Effective amplification of long targets from cloned inserts and human genomic DNA. *Proc. Natl Acad. Sci. USA* **91**: 5695–5699.
6. **Sharkey, D.J.** *et al.* (1994) Antibodies as thermolabile switches: high temperature triggering for the polymerase chain reaction. *BioTechnology* **12**: 506–510.
7. **Eckert, K.A. and Kunkel, T.A.** (1990) High fidelity DNA synthesis by the *Thermus aquaticus* DNA polymerase. *Nucleic Acids Res.* **18**: 3739–3744.

Further reading

Ballantyne, J., Sensabaugh, G. and Witkowski, J.A. (1989) DNA technology and forensic science, Banbury Report Number 32. Cold Spring Harbor Laboratory, New York.

Bej, A.K., Mahbubani, M.H. and Atlas, R.M. (1991) Amplification of nucleic acids by polymerase chain reaction (PCR) and other methods and their applications. *Crit. Rev. Biochem. Mol. Biol.* **26**: 301–334.

Dieffenbach, C.W. and Dvksler, G.S. (1995) *PCR Primer: a Laboratory Manual.* CSHL Press, Cold Spring Harbor, NY.

Gill, P., Jeffreys, A.J. and Werrett, D.J. (1985) Forensic applications of DNA fingerprints. *Nature* **318**: 577–579.

Innis, M.A., Yeland, D.H., Sninsky, J.J. and White, T.J. (1990) *PCR Protocols: a Guide to Methods and Applications.* Academic Press, San Diego, CA.

Kogan S.C., Doherty, M. and Gitschier, J. (1987) An improved method for prenatal diagnosis of genetic diseases by analysis of amplified DNA sequences. Application to hemophilia A. *N. Engl. J. Med.* **317**: 985.

Paabo, S., Higuchi, R.G. and Wilson, A.C. (1989) Ancient DNA and the polymerase chain reaction. The emerging field of molecular archaeology. *J. Biol. Chem.* **264**: 9709.

Reiss, J. and Cooper, D.N. (1990) Application of the polymerase chain reaction to the diagnosis of human genetic disease. *Hum. Genet.* **85**: 1–8.

Chapter 4

Application of PCR technology to human diseases

In addition to their central importance in understanding gene structure and function, PCR techniques are acquiring pivotal importance in disease diagnosis. Here we describe several typical such applications of this technology.

4.1 Genotyping using PCR technology

In Chapter 2 three classes of DNA markers detecting polymorphic variation were described: RFLPs, dynamic trinucleotide repeats and VNTRs. Trinucleotide repeats are just one class of microsatellites. Microsatellites are composed of di-, tri- and tetra-nucleotide repeats, for example $(CA)_n$ or $(CCA)_n$, where n can vary from 10 to more than 30. A typical example of a short microsatellite on the human Y chromosome was shown in *Figure 3.4*. In this particular individual there are 17 repeats of a CA sequence. Short microsatellites are simple to use in genotyping and, compared to Southern blotting (which is used for analysis of very long microsatellite repeats – see dynamic mutations in Chapter 2), they reduce the time taken to score a polymorphism from 1–2 weeks to less than 2 days. Microsatellite markers are often more informative than conventional RFLPs because they have many possible alleles. Also, they appear to be distributed randomly across the genome, unlike conventional VNTR markers which tend to be concentrated near the ends of chromosomes. Some microsatellites undergo uncontrolled expansion during the DNA synthesis phase of the cell cycle of meiosis or mitosis if the number of repeats passes a critical threshold. When the microsatellite is situated within a functionally important part of a gene then disease can result.

Markers defined using PCR require only minimal amounts of DNA and therefore microsatellites are increasingly being used in forensic medicine, paternity testing and medical diagnostics. An important use of micro-satellites is as an alternative to Southern transfer in forensic analysis of DNA, the great advantage being that tiny amounts of tissue, literally a blood spot, will give enough DNA to yield many genotypes. This makes it possible to identify an individual with a high degree of certainty.

Microsatellites are also widely used in gene mapping studies in conjunction with linkage analysis for locating the position of disease genes and susceptibility loci. They are also used in genetic diagnosis for a deductive technique called 'gene tracking'. Gene tracking is a rapid method of predictive and pre-natal genetic analysis. In order to use gene tracking, the diagnosis of the disease in the family must be certain and the location of the disease in relation to the microsatellite marker must also be known. Given a suitable family structure and relevant DNA samples, it is then possible to carry out a diagnosis without identifying the mutation segregating within the family (*Figure 4.1*).

4.2 Molecular analysis of clinical samples

Clinical molecular genetic laboratories are increasingly committed to testing patients for point mutations; a term commonly used to describe molecular scale lesions involving 1–100 nucleotides. This is as a direct consequence of the success of the Human Genome Project, which has led to the identification and characterization of the genetic basis of many inherited diseases. There are now a plethora of PCR-based techniques for detecting point mutations, but the choice of technique for any given

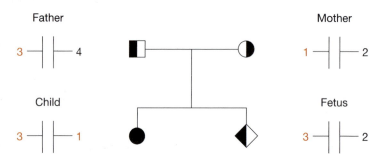

Figure 4.1. Gene tracking in a cystic fibrosis (CF) family. In order to have a child affected by CF, both parents must be carriers as CF is an autosomal recessive disease. The probability of each offspring of a carrier couple being affected is 1 in 4. In the family illustrated, who have requested pre-natal diagnosis, the CF mutations carried by the parents have not yet been characterized. Pre-natal diagnosis is still possible and has been carried out here using a microsatellite marker located within the CF gene. Testing with IVS17BTA has shown that both parents are informative (the alleles have been ranked 1 to 4 in order of decreasing molecular weight). The affected child has inherited allele '3' from their father and allele '1' from their mother. These alleles are inferred to be inherited along with uncharacterized mutations. The fetus inherits the high risk allele '3' from its father and the low risk allele '2' from the mother. This result is consistent with the fetus being a carrier of CF but unaffected. The probability of there being a diagnostic error due to recombination is very small (< 1%) since the marker is located within the gene and the recombination rate across the CF gene is < 0.5%.

disease depends on a number of factors, the most important of which is its molecular pathology. Since molecular pathology varies from disease to disease, so do the techniques that are applied. For instance, some diseases have 'common mutations' that are found in many unrelated cases and in other diseases the mutations are heterogeneous, that is they are different in unrelated cases; testing for diseases that have 'common mutations' employs a completely different set of techniques to those used in diseases with heterogeneous mutations. A third class of disease mutation analyzed by PCR is the dynamic mutations which, as explained previously, are caused by trinucleotide repeat expansions. These are dealt with by a different strategy again. Here we survey some of the PCR-based techniques and strategies used clinically to identify point mutations.

4.2.1 Testing for previously characterized mutations: the CF gene

Cystic fibrosis (CF) is a good example of a disease where many families carry the same mutation. CF is one of the most common autosomal recessive genetic diseases in Europeans. Patients typically suffer from persistent chest infections and impaired digestion. Eventually respiratory function becomes so impaired that patients need lung transplantation. The CF gene was cloned in 1989 [1–3] and over 900 different mutations have now been described worldwide.

One mutation, ΔF508, is extremely common in CF patients of European descent and is the commonest CF mutation in the world [4]. ΔF508 is a 3 bp deletion, which leads to the loss of a single phenylalanine codon from the CF gene. ΔF508 accounts for about 75% of all CF mutations in the UK [5]. Of the remaining mutations, the vast majority are exceedingly rare, although a few are either common in specific populations or found at low frequency in many populations. For instance the second most common mutation in the UK, G551D, comprises only 3% of the total. Using current techniques it would be uneconomic to test for all 900 mutations in clinical samples (it would be easier to simply sequence the whole gene in everyone!). However, by testing for a small subset of the mutations in any given population, it is possible to achieve an acceptable detection rate. For instance testing for the 20 most common CF mutations in the UK would detect about 85.5% of all CF mutations. This strategy of testing for a selection of the commonest CF mutations is the format that has been incorporated into clinically useful laboratory tests for CF. However, the molecular techniques that are employed vary between tests. Here we illustrate the workings behind two rapid and simple tests to detect the most common CF mutations in Europeans.

One of the CF mutation tests in routine clinical use utilizes a technique known as ARMS (amplification refractory mutation detection system) [6]. The ARMS technique exploits the fact that in order to be efficiently extended by DNA polymerases an oligonucleotide primer needs to have a perfectly matched 3′ end. Three PCR primers are used to fully genotype each mutation. One is a standard PCR primer, known as the common primer, that participates in both ARMS reactions. The remaining two PCR primers are specific, one for the normal sequence and one for the mutant sequence. The 3′ ends of the specific primers are designed to lie at the site of the mutation so that only the normal primer hybridizes perfectly to the normal sequence and *vice versa* for the mutant ARMS primer (*Figure 4.2*). In its simplest form, that is when testing for a single mutation, two ARMS PCR reactions are carried out, one in the presence of the common and normal ARMS primer, the second in the presence of the common and mutant ARMS primer. Amplification in the presence of the normal ARMS primer indicates the presence of the normal sequence in the sample, and amplification in the presence of the mutant ARMS primer indicates the presence of the mutant sequence. Consequently, all three possible genotypes for a given mutation, normal/normal, normal/ mutant and mutant/mutant, can be distinguished depending on whether one or both ARMS reactions amplify (*Figure 4.3*). A control amplification needs to be included within a single mutation ARMS assay such as this, because the absence of amplification is interpreted as meaningful. The control amplification should amplify regardless of whether the ARMS reaction amplifies and is usually from a completely unrelated part of the genome. The control PCR amplification should also amplify a larger DNA fragment than the ARMS reaction. This acts as a fail-safe for reactions that fail, either spontaneously or due to DNA degradation.

An alternative technology for testing for common CF mutations uses a method called the oligo ligation assay (OLA) [7,8]. Remarkably, this assay is able to genotype a DNA sample for 31 different CF mutations within a single reaction tube. There are three main stages in OLA analysis. The first step is PCR amplification of multiple DNA fragments from within the CF gene, corresponding to the regions where the mutations to be tested lie. The second stage involves multiple ligation reactions on the PCR

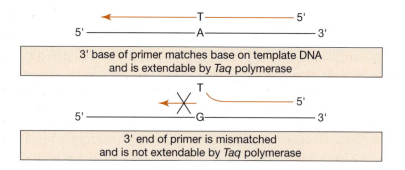

Figure 4.2. The principle of ARMS. Primers are designed so that their 3′ termini hybridize at the site of the mutated nucleotide. Imperfectly matched primer/ template complexes are incapable of extension by *Taq* polymerase.

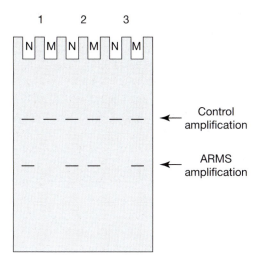

Figure 4.3. An ARMS reaction in its simplest form. Each sample is amplified in the presence of the normal specific ARMS primer (N) or the mutant specific ARMS primer (M) in conjunction with a common primer. Within each reaction a pair of primers for a control amplification has been added to ensure that the PCR reaction was capable of amplification. Lane 1 shows the pattern expected from a normal homozygoyte, lane 2 a normal/mutant heterozygote and lane 3 a mutant homozygote.

amplified DNA using different sets of oligonucleotides. OLA uses three oligonucleotides to genotype each specific locus on a PCR amplified fragment. A pair of oligonucleotides acts as differential probes; one is specific for the normal sequence and one for the mutant sequence. They are designed so that their 3' nucleotides hybridize perfectly to either the normal or mutant sequence. Each probe oligo carries at its 5' end a synthetic retarding group (pentaethyleneoxide) that confers a distinct mobility, thus allowing the normal and mutant ligation products to be distinguished after gel electrophoresis. The third oligonucleotide acts as a reporter molecule and carries one of three different fluorescent labels at its 3' end. The different fluorescent labels permit the OLA fragments to be detected by both size and fluorescent label, thus permitting the detection of many mutations in a single reaction. The 5' end of the reporter oligonucleotide is designed to hybridize exactly adjacent to the 3' end of the probe oligos. In order for a ligation reaction to occur efficiently the 3' and 5' termini of two DNA molecules must be perfectly hybridized. If the probe oligonucleotide hybridizes perfectly to the test DNA sample, the 3' terminus of the probe and 5' terminus of the reporter oligo can be ligated. However, a mismatched base at the 3' end of the probe oligo prevents ligation to the reporter oligo (*Figure 4.4*). A thermostable ligase is employed in the CF OLA assay, permitting multiple cycles of annealing, ligation and denaturation without addition of fresh enzyme between cycles. Finally the ligation products are electrophoresed on denaturing polyacrylamide gels on a fluorescent DNA analyzer which can separate them both by mobility and fluorescent label (*Figure 4.5*).

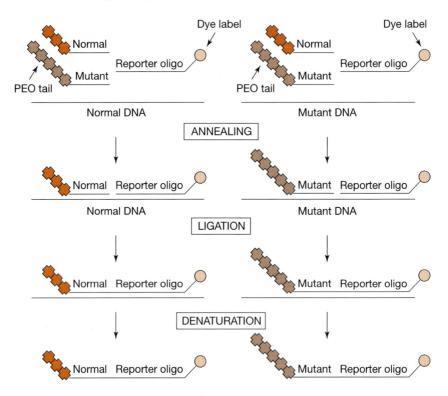

Figure 4.4. Principle of the oligonucleotide ligation assay (OLA). Ligation is carried out in cycles on PCR amplified DNA; a single cycle for a single locus is illustrated. Ligation consists of the three steps, annealing, ligation and denaturation. Note that the dye labeled reporter oligo is the same for both normal and mutant alleles for each locus but that the normal and mutant OLA products are differentiated by the lengths of their PEO mobility modifying tails.

4.2.2 Testing for uncharacterized mutations: the Nf2 gene

Neurofibromatosis type 2 (Nf2) is an example of a disease where there is strong mutational heterogeneity, that is where mutations are different between families. About 65% of mutations in Nf2 are point mutations [9], the remainder being large deletions of all or part of the gene. Nf2 is an inherited cancer syndrome, its hallmark being the development of bilateral tumors on the sheath surrounding the acoustic nerve. Nf2 affects about one in 37 000 adults in the UK and is inherited in an autosomal dominant fashion [10]. Since in Nf2, point mutations are equally likely to occur anywhere within the gene, the whole of the coding sequence including splicing signals needs to be tested. Consequently a mutation scanning strategy is used to examine the whole gene quickly but imperfectly. A mutation scan indicates which DNA fragments from an individual differ from a normal control and thus need sequencing. However, such scans do

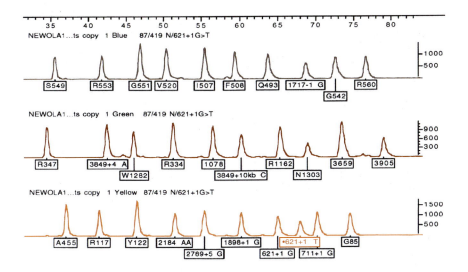

Figure 4.5. CF OLA results. A single sample generates three traces corresponding to the different fluorescent dyes used to label the reporter oligos. Each trace has specific peaks with mobilities corresponding to either normal or mutant OLA products at the tested loci. These peaks have been labeled with the appropriate loci by proprietary software. On the lower (yellow label) trace an extra, mutant OLA, peak has been detected. This corresponds to the mutation 621+1G>T confirming that this patient is a carrier of this mutation.

not reveal all possible mutations and most are incapable of distinguishing between neutral polymorphisms and pathogenic mutations. There are many different mutation scanning techniques available, each with their own attendant pros and cons. The method used for any particular disease will depend on a compromise between throughput and detection efficiency. A mutation scan typically examines the gene in several small fragments, the number of fragments depending on both the size and structure of the gene and the technique used. Here we describe two scanning methods that have been applied to Nf2: single strand conformation polymorphism (SSCP)/heteroduplex analysis and chemical cleavage of mismatches (CCM) analysis. In SSCP/heteroduplex analysis a fragment from the gene of interest, typically an exon, is PCR amplified. The sample is then mixed with formamide and heat denatured in order to create single-stranded DNA (ssDNA). The denatured sample is snap-chilled on ice and electrophoresed at low temperature on a polyacrylamide gel. As the ssDNA begins to migrate into the polyacrylamide gel matrix the denaturant effect of the formamide is quickly lost. Some of the DNA re-anneals back to form double-stranded DNA (dsDNA), but a sizeable proportion remains single-stranded. The ssDNA forms a three-dimensional structure (conformation) that is mediated by base pairing in regions of self-complementarity. This structure is sequence dependent and single

base changes, like those produced by mutations, can alter the conformation of the structure. The migration rate of the ssDNA is related to its
shape (*Figure 4.6*). Hence, point mutations can give rise to mobility shifts
[11]. SSCP analysis is not a perfect mutation scanning method. It is most
effective on small fragments of about 150 bp in size, when about 90% of
all mutations are detected [12].

The small amount of dsDNA that forms during the initial electrophoresis can also aid in the identification of point mutations by forming
heteroduplexes. Heteroduplexes are hybrid DNA molecules which are
formed when a PCR product containing two separate alleles is denatured
and spontaneously re-anneal (*Figure 4.7*). The corresponding perfectly
matched dsDNA molecules are called homoduplexes. In the case of a
heterozygote for a point mutation, half the dsDNA can be present in the
form of heteroduplexes. The mismatched base or bases present in the
heteroduplexes disrupts the overall structure of the DNA and is thought
to induce bends or regions of rigidity which affect the overall mobility of
the heteroduplex DNA. Consequently, samples that are heterozygous for
mutations are capable of giving rise to heteroduplex mobility shifts on gel
electrophoresis [13]. As with SSCP analysis, heteroduplex analysis is not a
foolproof mutation detection method. However, by combining SSCP and
heteroduplex analysis, point mutation detection rates approaching 90%
on a 300 bp fragment can be achieved. *Figure 4.8* gives an example of an
SSCP/heteroduplex gel with some NF2 mutations present.

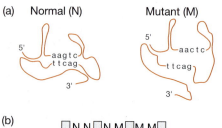

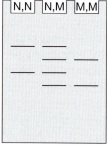

Figure 4.6. SSCP analysis.
(a) Two ssDNA molecules of
different sequence (normal and
mutant) under non-denaturing
conditions are illustrated. Each
differs in sequence by a single
nucleotide. The normal DNA has
a region of self-complementarity
which has been disrupted in the
mutant sequence. The region of
self-complementarity participates
in the folding of the DNA,
consequently the shape adopted
by the normal and mutant DNA
is different. (b) The difference in
shape results in different
mobilities after polyacrylamide
gel electrophoresis. Note that two
bands are typically observed for a
single sequence corresponding to
the two DNA strands, resulting
in four bands in the normal
mutant heterozygote.

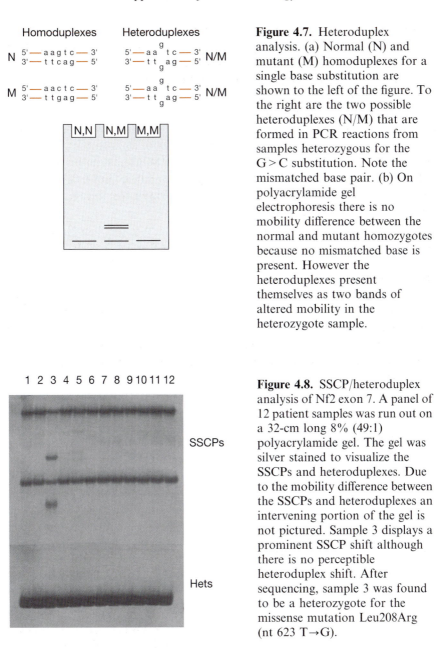

Homoduplexes   Heteroduplexes

N $\begin{array}{l} 5'-aagtc-3' \\ 3'-ttcag-5' \end{array}$ $\begin{array}{l} 5'-aa\overset{g}{}tc-3' \\ 3'-tt\underset{g}{}ag-5' \end{array}$ N/M

M $\begin{array}{l} 5'-aactc-3' \\ 3'-ttgag-5' \end{array}$ $\begin{array}{l} 5'-aa\overset{g}{}tc-3' \\ 3'-tt\underset{g}{}ag-5' \end{array}$ N/M

Figure 4.7. Heteroduplex analysis. (a) Normal (N) and mutant (M) homoduplexes for a single base substitution are shown to the left of the figure. To the right are the two possible heteroduplexes (N/M) that are formed in PCR reactions from samples heterozygous for the G > C substitution. Note the mismatched base pair. (b) On polyacrylamide gel electrophoresis there is no mobility difference between the normal and mutant homozygotes because no mismatched base is present. However the heteroduplexes present themselves as two bands of altered mobility in the heterozygote sample.

Figure 4.8. SSCP/heteroduplex analysis of Nf2 exon 7. A panel of 12 patient samples was run out on a 32-cm long 8% (49:1) polyacrylamide gel. The gel was silver stained to visualize the SSCPs and heteroduplexes. Due to the mobility difference between the SSCPs and heteroduplexes an intervening portion of the gel is not pictured. Sample 3 displays a prominent SSCP shift although there is no perceptible heteroduplex shift. After sequencing, sample 3 was found to be a heterozygote for the missense mutation Leu208Arg (nt 623 T→G).

In contrast to ARMS and OLA, SSCP/heteroduplex analysis is a non-specific technique and cannot be used alone to define mutations. A mobility shift on a gel does not give any information about the underlying difference, which may be due to either a polymorphism or a pathogenic mutation. Consequently sequencing is necessary to determine the underlying cause of all mobility shifts observed.

CCM analysis is a method of mutation scanning for point mutations that relies on the chemical modification and cleavage of mismatched bases in heteroduplex DNA [14]. It uses the principle of Maxam and Gilbert DNA sequencing whereby specific bases are modified followed by cleavage of the ribose-phosphate backbone at the site of modification (see Section 3.1.1). The chemically modified mismatched bases are then cleaved using piperidine and the labeled DNA run out on denaturing polyacrylamide gels. The presence of a mutation leads to the production of two DNA fragments of lower molecular weight than the uncleaved product (*Figure 4.9*). Provided that the DNA is labeled appropriately, it is theoretically possible to detect all possible mismatches and thus all point mutations (*Figures 4.9* and *4.10*).

4.2.3 Testing for dynamic mutations by PCR

Dynamic mutations are a class of disease mutation caused by unstable trinucleotide repeats (see also Section 2.4.3). As their name implies they are highly mutable, particularly in their expanded state, when they frequently undergo intergenerational changes in size. Dynamic mutations can be sited within different parts of a gene, for instance within the coding sequence (Huntington's disease [HD]), within an intron (Friedreich's ataxia), within the 5'UTR (fragile X syndrome) or within the 3'UTR (myotonic dystrophy). Dynamic mutations do not always act in the same way; some obliterate gene expression, such as Friedreich's ataxia and fragile X syndrome, whereas others lead to a gain of function, such as HD. Some expansions of dynamic mutations can be very large and we have previously dealt with their analysis by Southern blotting in Chapter 2. Smaller expansions pose a different set of problems but they can be analyzed by PCR amplification of the repeat followed by denaturing polyacrylamide gel electrophoresis.

HD is a good example of a dynamic mutation that is mainly analyzed by PCR. HD is inherited in an autosomal dominant fashion and is caused by the expansion of a CAG_n repeat within exon 1 of the Huntingtin gene [15]. Three to seven individuals per 100 000 within Western Europe are affected. Patients with HD first display symptoms in middle age, generally beginning with subtle changes in neurological function, but eventually progressing to an involuntary movement disorder called a chorea. Neurological function progressively deteriorates, eventually leading to death. The CAG_n repeat is polymorphic in normal individuals, varying in size from 9 to 35 repeats. HD-affected patients have expansions ranging from 40 to 121 repeats. The normal and affected ranges are non-overlapping, although there are rare instances of what are termed intermediate alleles which fall in-between the normal and expanded ranges. Some patients with intermediate alleles may become affected, while others do not. Testing for HD involves PCR amplification of the

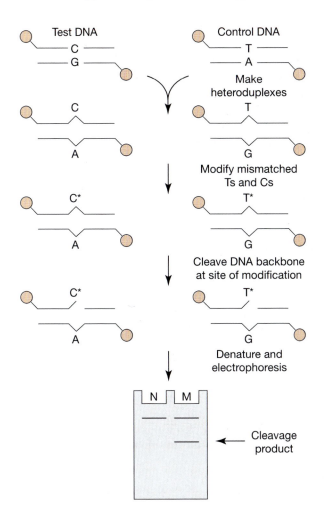

Figure 4.9. Chemical cleavage of mismatch (CCM) analysis. A single base substitution is present in the test DNA. The control DNA and test DNA are both PCR amplified using fluorescently labeled primers. Heteroduplexes are then made by mixing the test and control DNA together followed by a single cycle of denaturation and annealing. The DNA is then split, half being treated with hydroxylamine which modifies mismatched cytosines and the other half with either OsO_4 or $KMnO_4$ which modify mismatched thymines. The phosphate backbone of the DNA is then cleaved at the site of mismatch using piperidine. Finally the treated DNA samples are run out on a denaturing polyacrylamide gel. The presence of a novel band of higher mobility than the full length PCR product indicates that there is a mutation present in the sample.

CAG_n repeat by primers designed to hybridize to the flanking sequence. Due to the high GC content of the repeat and its consequently high melting temperature, the HD CAG_n repeat is a difficult template to amplify efficiently. After amplification the PCR product is analyzed by

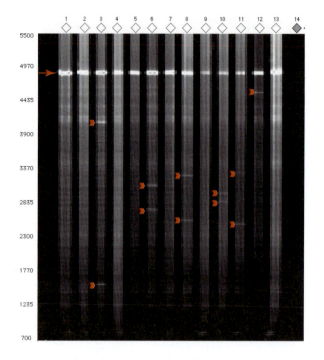

Figure 4.10. Gel image from fluorescent CCM analysis. Samples heterozygous for mutations in the Nf2 gene were PCR amplified and labelled with the fluorescent dye label 'TAMRA'. Mismatched cytosine bases were modified with hydroxylamine and then cleaved with piperidine before electrophoresis. The full length labeled product is marked with a yellow arrow. Cleavage products, marked with orange chevrons, can clearly be seen in lanes 3, 6, 8, 10, 11 and 12.

denaturing polyacrylamide gel electrophoresis (PAGE). Samples are run alongside molecular weight standards so that the size of each allele can be accurately measured (*Figure 4.11*). Patients with two alleles within the normal range cannot have or develop HD. Patients with alleles in the expanded range are either affected or will become affected. A subset of patients only have a single allele within the normal range. This is almost always due to the fact that the patient is homozygous for the allele. However it is not always possible to completely exclude the possibility that the patient has a single allele within the normal range and an expansion that is too large to be resolved by PCR and PAGE. In these cases Southern blotting is sometimes necessary to exclude this possibility.

4.3 Diagnosing infectious diseases by PCR

Another application of PCR for diagnostic purposes is in the identification of DNA brought into the patient's tissues by an infectious agent. This technique offers several advantages over the more commonly used serological and biological procedures. Because PCR is so sensitive it

M 1 2 M 3 4 5 M

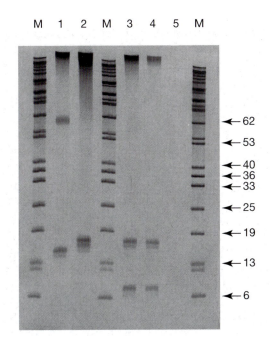

←62

←53

←40
←36
←33

←25

←19

←13

←6

Figure 4.11. Mutation analysis in Huntington's disease. Samples were PCR amplified with primers flanking the HD CAG_n repeat and electrophoresed on a 6% (19:1) denaturing polyacrylamide gel. DNA was then visualized using silver staining. Lanes containing molecular weight standards are marked with 'M'. The number of CAG_n repeats corresponding to fragments in the molecular weight marker are given on the right. Sample 1 is from an individual diagnosed with HD. An expansion of approximately 58 repeats can clearly be seen. Samples 2, 3 and 4 are from patients requesting predictive tests. Samples 3 and 4 both have two alleles within the normal range thus excluding a diagnosis of HD. Sample 2 however only has a single allele within the normal range. Although this individual is likely to be homozygous for the normal allele, it is not possible to unequivocally exclude the presence of a large expansion refractory to PCR. Sample 5 is a no DNA control. Photograph courtesy of Alan Dodge, St. Mary's Hospital, Manchester.

allows an earlier diagnosis to be made. In addition, when direct identification of a viable infectious agent is not possible, detection of that agent's DNA is considered good evidence of its viability and potential to self-replicate within the host. In contrast, serology identifies antigens that may have been released into the bloodstream. These may be no longer associated with viable viral or bacterial particles. PCR analysis can also be of particular value with immunosuppressed patients, where detection of specific antibodies directed against the infecting agent is difficult. It is similarly useful in analyzing fetal blood, as not all maternal antibodies can cross the placental barrier and the fetus itself synthesizes extremely small amounts of antibodies. In patients infected with hepatitis B virus (HBV), clinical evaluation, paralleled by ordinary serological data, is sufficient for a diagnosis to be made. The situation is complicated in the case of a chronic carrier. Direct identification of the HBV DNA in the patient's blood proves that the infectious state is still present. Identification of the 'e' antigen (a protein present in the core of the viral particle), although of the same significance, is not as sensitive. Proof that the viral infection is still active is also of prognostic importance, because it is associated with development and persistence of chronic hepatic damage (chronic hepatic cirrhosis or hepatocarcinoma). In the case of congenital toxoplasmosis, serological analysis of fetal blood is virtually ineffective and here PCR is of great value. Fetal infection causes a number of pathological situations depending on the period of pregnancy during which infection has occurred. These range from oculopathy to gross malformations in the central nervous system with resultant irreversible brain damage. The small size of the fetal blood specimen, and the low concentration of the microbial DNA, make the sensitivity of the PCR reaction a great advantage. Pharmacological therapy can give good results only if an early diagnosis, prior to irreversible brain damage or malformations, can be made. Again, serology is of no use in immunodepressed acquired immuno deficiency syndrome (AIDS) patients, where toxoplasmosis is among the most common opportunistic infections. Other infectious agents for which PCR analysis is very valuable are cytomegalovirus (CMV), chlamydiae, and herpes virus (HV). These viruses affect the infant at the time of birth. The conventional analysis methods are, in the case of HV or chlamydial infections, immunohistochemical characterization of uterine cervix cells and, for CMV, isolation from cultured neonatal urinary sediment cells. Such approaches are time-consuming, while PCR analysis allows rapid identification even if minimal amounts of the virus are present.

References

1. **Rommens, J.M., Ianuzzi, M.C., Kerem, B.** *et al.* (1989) Identification of the cystic fibrosis gene: chromosome walking and jumping. *Science* **245**: 1059–1065.
2. **Riordan, J.R., Rommens, J.M., Kerem, B.** *et al.* (1989) Identification of the cystic fibrosis gene: cloning and characterization of complementary DNA. *Science* **245**: 1066–1073.
3. **Kerem, B., Rommens, J.M., Buchanan, J.A.** *et al.* (1989) Identification of the cystic fibrosis gene: genetic analysis. *Science* **245**: 1073–1080.
4. **Kazazian, H.H. and The Cystic Fibrosis Genetic Analysis Consortium** (1994) Population variation of common cystic fibrosis mutations. *Hum. Mutat.* **4**: 167–177.
5. **Schwarz, M.J., Malone, G.M., Haworth, A.** *et al.* (1995) Cystic fibrosis mutation analysis: report from 22 UK regional genetics laboratories. *Hum. Mutat.* **6**: 326–333.
6. **Newton, C.R., Graham, A., Heptinstall, L.E.** *et al.* (1989) Analysis of any point mutation in DNA: the amplification refractory mutation system. *Nucleic Acids Res.* **17**: 2503–2516.

7. **Landegren, U., Kaiser, R., Sanders, J. and Hood, L.** (1988) A ligase-mediated gene detection technique. *Science* **241**: 1077–1080.
8. **Brinson, E.C., Adriano, T., Bloch, W.** *et al.* (1997) Introduction to PCR/OLA/ SCS, a multiplex DNA test, and its application to cystic fibrosis. *Genet. Test.* **1**: 61–68.
9. **MacCollin, M., Ramesh, V., Jacoby, L.B.** *et al.* (1994) Mutational analysis of patients with neurofibromatosis 2. *Am. J. Hum. Genet.* **55**: 314–320.
10. **Evans, D.G., Huson, S.M., Donnai, D.** *et al.* (1992) A genetic study of type 2 neurofibromatosis in the United Kingdom. I. Prevalence, mutation rate, fitness and confirmation of maternal transmission effect on severity. *J. Med. Genet.* **29**: 841–846.
11. **Orita, M., Suzuki, Y., Sekiya, T. and Hayashi, K.** (1989) Rapid and sensitive detection of point mutations and DNA polymorphisms using the polymerase chain reaction. *Genomics* **5**: 874–879.
12. **Sheffield, V.C., Beck, J.S., Kwitek, A.E., Sandstrom, D.W. and Stone, E.M.** (1993) The sensitivity of single-strand conformation polymorphism analysis for the detection of single base substitutions. *Genomics* **16**: 325–332.
13. **Keen, J., Lester, D., Inglehearn, D., Curtis, A. and Bhattacharya, S.** (1991) Rapid detection of single base mismatches as heteroduplexes on Hydrolink gels. *Trends Genet.* **7**: 5.
14. **Cotton, R.G., Rodrigues, N.R. and Campbell, R.D.** (1988) Reactivity of cytosine and thymine in single-base-pair mismatches with hydroxylamine and osmium tetroxide and its application to the study of mutations. *Proc. Natl Acad. Sci. USA* **85**: 4397–4401.
15. **Huntington's Disease Collaborative Research Group** (1993) A novel gene containing a trinucleotide repeat that is expanded and unstable on Huntington's disease chromosomes. *Cell* **72**: 971–983.

Further reading

Strachan, T. and Read, A.P. (1999) *Human Molecular Genetics,* 2nd Edn. BIOS Scientific Publishers, Oxford.
Cotton, R.G.H., Edkins, E. and Forrest, S. (1998) *Mutation Detection: a Practical Approach.* IRL Press, Oxford.
Elles, R. (ed.) (1996) *Molecular Diagnosis of Genetic Diseases.* Humana Press, Totowa, NJ.
Taylor, G.R. (ed.) (1997) *Laboratory Methods for the Detection of Mutations and Polymorphisms in DNA.* CRC Press, Boca Raton, FL.

Chapter 5

Cloning in *E. coli* I: vectors and genomic libraries

While PCR offers the possibility of amplifying specific segments of DNA in the test tube, there are many purposes for which gene cloning remains essential. These include the initial isolation of a gene, its propagation in the laboratory and the production of its protein product. The two methods, PCR and gene cloning, should be viewed as complementary techniques.

5.1 Cloning and cloning vectors

Cloning constitutes a one-step purification of such enormous power that single genes, comprising only one part in one million of total human genomic DNA, can readily be isolated. In gene cloning, a replication-competent DNA molecule, termed the vector, is introduced into a suitable host cell. Here we describe cloning in the bacterial host *E. coli*, which is the general workhorse of genetic engineering. Cloning involves three basic steps: (a) the vector DNA is cleaved with one or more restriction enzymes; (b) the DNA to be cloned, the target or insert, is joined to the vector, generating a recombinant molecule; and (c) the recombinant DNA molecules are introduced into the host bacterial cell, which is said to be transformed by the introduction of the vector molecule.

If a mixture of recombinant molecules is introduced into a host cell population at low efficiency, so that most cells receive only one DNA molecule, then each colony of cells that grows up on an agar plate will contain a different recombinant DNA molecule. Each colony can be picked separately, or cloned, and the particular recombinant molecule that it contains can be isolated, identified and amplified.

A cloning vector is necessary primarily as a carrier of the cloned gene. Without a vector, a DNA molecule introduced into a cell would be diluted out by cell division and eventually lost. There are a number of features common to all cloning vectors. First, the vector must be able to replicate in the host cell. Hence it must have a replication origin, a sequence element that is recognized by the host cell's replication machinery, and that directs amplification of the vector. Second, transformation is a somewhat inefficient process; only a minority of cells in the population take up and retain the exogenous DNA. Therefore a further requirement is that the vector contains a selectable marker, so that the sub-population of bacteria containing it can be isolated.

Some vectors are based upon extrachromosomal circular DNA molecules called plasmids. These cloning vectors derive from naturally occurring drug-resistance plasmids. An antibiotic resistance gene normally encodes an enzyme which modifies or breaks down the antibiotic; for example, ampicillin resistance is conferred by a β-lactamase which degrades the drug. Clones of transformant cells containing a drug resistance plasmid are selected by their ability to grow in the presence of a concentration of the antibiotic which prevents the growth of, or is lethal to, non-transformed cells. There are also vectors based upon bacterial DNA viruses, such as bacteriophage. When a phage particle infects a cell, circular areas of dead bacteria called plaques are formed in the 'lawn' of healthy cells. Each plaque contains the viral progeny of a single infectious particle, and the virus can be picked from it and amplified by growth in a fresh host cell.

5.2 Plasmid vectors and their use in manipulating DNA

The three basic steps in any cloning procedure – constructing the recombinant DNA molecule, introducing it into the bacterium and selecting a transformed cell containing the desired recombinant – are performed most simply using plasmid vectors. Plasmids are, for example, invariably the vector class of choice for isolating sub-regions from segments of DNA that have already been purified by gene cloning (to yield 'sub-clones') or to clone PCR products.

5.2.1 Multi-purpose vectors

Modern plasmid vectors evolved from simpler plasmids that contain little more than a replication origin and a method of selection, by the addition of sequence features that extend their usefulness. pBluescript is a commonly used vector and it illustrates some of the improvements that have been made over its 'grandparent' (a plasmid called pBR322) and its 'parent' (pUC19). pBluescript carries a short DNA sequence containing many closely spaced restriction enzyme cleavage sites (*Figure 5.1*). Such a multi-cloning site (MCS) or polylinker is constructed by chemical synthesis. An MCS increases the number of potential cloning strategies available, by extending the range of enzymes that can be used to generate a restriction fragment suitable for cloning. By combining them within an MCS the restriction sites are made contiguous, so that any two sites within it can be cleaved simultaneously without excising vector sequences.

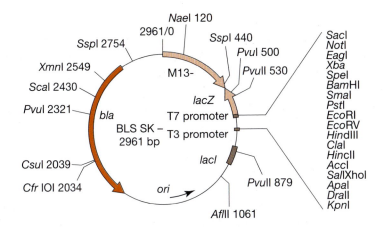

Figure 5.1. The pBluescript plasmid vector. The pBluescript plasmid is 2961 bp long and contains two origins of replication: the *E. coli ori* and the M13 *ori*. M13 is a filamentous phage with a ssDNA genome. In the presence of phage proteins, encoded by a 'helper phage', pBluescript can be propagated and isolated as ssDNA. The orientation of the M13 origin dictates which of the two parental strands (+ or −) will be copied and incorporated into phage particles. The (−) version is presented here. The MCS incorporates sites for several useful enzymes. In the SK version the MCS starts with *Sac*I and ends with *Kpn*I but in the KS version it carries the same sites in the reverse order.

In addition to facilitating the cloning procedures, the MCS in pBluescript lies within a short region of DNA sequence derived from the 5′ end of the *lac* operon of *E. coli* (*Figure 5.2*). This region contains the promoter of *lacZ*, the *E. coli* gene that encodes β-galactosidase (β-gal), and a small part of the β-gal coding region that encodes a sequence known as the α-peptide. The α-peptide will combine with a mutant β-gal protein which lacks the normal N-terminus, to generate a functional enzyme molecule. When an *E. coli* cell containing such a mutant *lacZ* gene within its genome is transformed with a pBluescript vector, enzymatically active β-gal accumulates. If a chromogenic β-gal substrate is incorporated in the agar plates, a blue product accumulates in the colonies, harboring the vector with the intact MCS–*lacZ* gene fusion. Normally, any fragment of DNA inserted into the MCS will, at some point along its length, contain a termination codon in the same reading frame as the α-peptide. This will prevent synthesis of the α-peptide. Hence, bacterial colonies containing recombinant DNA molecules are colorless and can be easily identified from the 'background' of blue, non-recombinant clones. This feature of the vector therefore provides a visual selection method, to indicate the presence of a cloned insert in the MCS.

Flanking the pBluescript MCS there are sequences of the promoters of the RNA polymerases derived from bacteriophages T3 and T7. By cloning a gene into the MCS and copying with the appropriate polymerase *in vitro* (i.e. either T3 or T7 polymerase), RNA transcripts which have the same

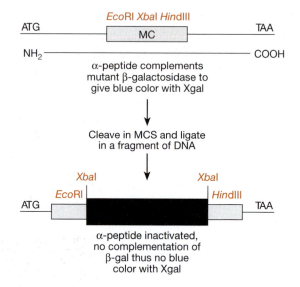

Figure 5.2. The principle of *lacZ* selection. An MCS of this form is present in many plasmid and phage vectors. As well as allowing application of a simple blue–white test for the presence of an insert, the *lac* fragment provides a strong bacterial promoter and Shine–Delgarno signal. Thus, genes inserted into such an MCS in the correct orientation and in frame with the β-gal ATG initiation codon are expressed in *E. coli* at a high level. If the resulting fusion protein is stable, it can be isolated and analyzed further or utilized. Redrawn from Williams and Patient (1989) with permission from Oxford University Press.

sequence as the gene ('sense' RNA) or the complementary sequence ('anti-sense' RNA) are generated (*Figure 5.3*). Sense and anti-sense RNAs have many uses. For example, sense RNA can be used to direct synthesis of the cognate protein, using an *in vitro* translation system derived from rabbit

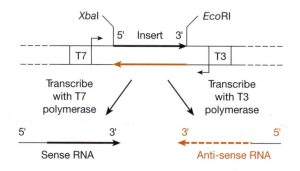

Figure 5.3. Sense and anti-sense RNA production using bacteriophage polymerases. The coding strand of the insert is shown as a solid black arrow. When the insert DNA is copied with T7 polymerase, the transcript will have the sequence of the coding strand of the gene from which it derives, i.e. if translated it would yield the cognate protein. Hence, it is called 'sense' RNA. When copied with T3 polymerase, the transcript will have the sequence of the non-coding strand of the gene and is therefore called 'anti-sense' RNA.

reticulocytes. Anti-sense RNA can be used as a hybridization probe to detect transcripts of the gene by Northern blotting.

pBluescript also incorporates the replication origin of the bacteriophage M13. This short fragment of DNA, responsible for directing the replication of the single-stranded genome of the M13 phage, allows the easy preparation of single-stranded pBluescript DNA together with the insert cloned into it. When bacteria containing such a vector are infected with 'helper' M13 phage, which encode the proteins needed for viral replication, single-stranded DNA copies of the plasmid vector are produced and packaged into pseudo-viral particles. Single-stranded DNA can be isolated from these particles by lysing them and purifying the DNA away from the phage proteins. The M13 replication origin works uni-directionally and, depending upon its orientation within the vector, it will allow isolation of either strand of the pBluescript DNA. These can then be used, for example, as templates for DNA sequence analysis.

5.2.2 DNA sub-cloning in a plasmid vector

DNA sub-cloning into a plasmid vector usually involves the following steps (*Figure 5.4*):

(i)   the target to be cloned is cleaved from the source DNA molecule;
(ii) the vector molecule is cleaved in its MCS;
(iii) the target and vector molecules are joined;
(iv) *E. coli* cells are transformed using the recombinant vector.

For the first step, cleavage of the target sequence from the source, either a single restriction enzyme or a pair of enzymes is used. If there are no restriction enzyme sites at the required positions within the starting molecule, then PCR is often used to generate a product with the desired termini. By incorporating restriction sites into the PCR primers (at the 5′ end of the primers, where pairing with the template is not critical), total control can be obtained over the nature of the final product. The target is usually purified on an agarose or polyacrylamide gel to remove unwanted molecules, which would interfere with cloning.

Since plasmids are circular, the second step, cleaving the vector within the MCS, is often referred to as linearization. This is done with an enzyme, or with two enzymes, that generate ends compatible with the termini on the target. In the simplest case the ends will be blunt. Any blunt-ended molecule will join to any other blunt-ended molecule. However, the efficiency of joining is significantly lower for blunt than for staggered ends, because the annealing of cohesive termini greatly favors their joining. If possible, therefore, enzymes which generate sticky ends are used.

The enzyme used to generate the target need not necessarily be the same enzyme used to linearize the vector, but the enzymes must produce compatible overhanging termini. For example, *Sau*3A recognizes the

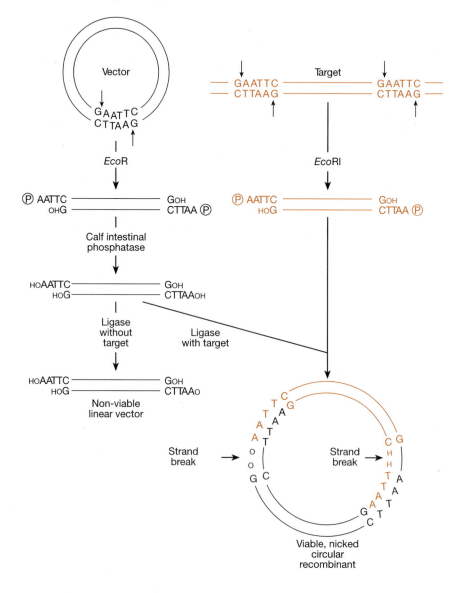

Figure 5.4. A typical sub-cloning experiment. Redrawn from Williams and Patient (1989) with permission from Oxford University Press.

sequence GATC and *Bam*HI recognizes the sequence GGATCC but both enzymes generate a 5′ overhanging terminus with the same sequence, GATC. Hence a *Sau*3A fragment can be cloned into a *Bam*HI site. This enzyme combination is often used, for example, to generate genomic DNA libraries (see Section 5.4.1; note, the resulting junctions, i.e. *Sau*3A to *Bam*HI, will only be cleavable by *Bam*HI if the particular *Sau*3A site involved has a C residue as its 3′ neighboring residue).

If a single enzyme, or two enzymes that generate the same sticky end, are used for cloning then the target fragment may come to lie in either orientation. If two different sticky-end enzymes are used to generate the target, then the fragment will have a fixed orientation with respect to the vector molecule. This is referred to as a 'forced' or a 'directional' cloning.

The third step, joining the target and vector molecules, uses the enzyme T4 DNA ligase. The desired reaction is a two-step ligation, in which a molecule of target joins first to one end, and then to the other end, of a vector molecule. This will generate a circular molecule containing the target DNA. Linear DNA has a transformation efficiency several orders of magnitude lower than circular DNA. Hence recombinant molecules are cloned much more frequently than linear vector molecules. However, there are a number of undesirable competing ligation reactions, such as end-to-end joining of vector or target molecules and recircularization of the vector without a target molecule. The competing reactions can be controlled to an extent by varying the ratios of enzyme and target, but recircularization of the linearized vector is a very frequent event (because the ends are held in such close relative proximity) and cannot be completely eliminated in this way. Vector recircularization generates a background of non-transformant clones that will greatly exceed the number of clones containing recombinant DNA molecules. A number of tricks are available to minimize the problem.

One of these tricks is to accept the high background but to identify recombinants by cloning into an MCS, such as that of the pBluescript vectors, where β-gal screening selection can be performed (*Figure 5.2*). Alternatively, if two restriction enzymes which generate different cohesive termini are used to cleave the vector for cloning (that is if forced cloning is performed) then the background will be greatly reduced, because the vector cannot recircularize. If neither of these options is practicable, then the background can be reduced by treating the vector with a phosphatase prior to ligation (*Figure 5.4*). Phosphatase removes the 5′ terminal phosphate groups from the termini. T4 DNA ligase requires a 5′ phosphate in order to join two molecules, so vector molecules cannot recircularize. However, target molecules can ligate to vector molecules to form a circular molecule. Those vector molecules which recircularize by joining with a target molecule are able to transform *E. coli*. Although only one covalent bond is formed at each junction of vector with target (using the phosphate donated by the target), this is sufficient to keep the vector and target molecules attached and the gaps are repaired once the DNA is introduced into *E. coli*.

The fourth step, transformation, can be done in a number of different ways. Most commonly, bacterial membranes are rendered temporarily permeable to DNA by exposure to divalent cations or by a brief electric shock (electroporation). The bacteria are then placed under conditions selective for the growth of transformant cells, normally by plating on an agar plate in the presence of an antibiotic. If the target is pure, and the procedure used to select against vector recircularization is successful, most or all of the bacterial clones will contain the desired sequence. However, this is normally checked by growing small bacterial cultures ('mini-preparations') derived from individual colonies. The cells are lysed, plasmid DNA is extracted and a restriction enzyme cleavage map is determined to verify that the DNA has the expected structure. Only then is a large-scale plasmid preparation made.

There are many variations to the above scheme and two very important procedures that combine PCR technology with plasmid cloning are the direct cloning of PCR products and site-directed mutagenesis.

5.2.3 Direct cloning of PCR products

When the sequence of a DNA fragment is known it can be amplified by PCR and then cloned. This can be done by incorporating restriction enzyme cleavage sites into the PCR primers, as described above, but the PCR product can also be cloned by direct ligation. Direct cloning of PCR products relies on the fact that thermostable polymerases used for PCR incorporate an additional A residue at the 3' end of the newly synthesized DNA chains. The vector DNA, now usually purchased commercially, is a linearized molecule with the extremities modified by the presence of an extra T residue. After the PCR reaction, the PCR product can be directly ligated into such a plasmid molecule, without the need for further enzymatic manipulation. The presence of the additional T residue is also of benefit because it inhibits recircularization of the vector. Recently the ability of the enyme topoisomerase I (Topo-I) to join DNA ends has been exploited to increase the speed and efficiency of cloning PCR products even further. Here linearized vector molecules carrying the extra T residue are sold as a pre-associated complex with Topo-I. When the vector and target molecules transiently associate, through their A- and T-modified termini, the topoisomerase seals them together.

5.2.4 Site-directed mutagenesis

Once a gene has been isolated by cloning, it is sometimes necessary to modify its sequence in some way, perhaps to change the biological properties of the encoded protein or to optimize its expression in a host organism. This is achieved by site-directed mutagenesis. A method in common use is shown in *Figure 5.5*. It relies on a series of PCR steps, one of which uses primers that incorporate the changes to be made [2]. These mismatches will partially destabilize the annealed product but, provided there is a sufficient region of perfectly homologous sequence flanking the mismatch, a hybrid will be formed which can be recovered by cloning in *E. coli*.

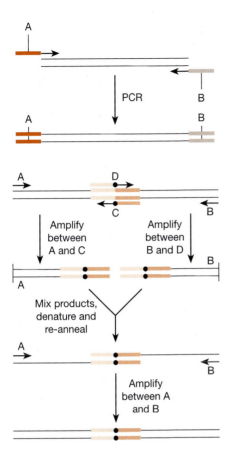

Figure 5.5. Site-directed mutagenesis. The region encompassing the site to be mutagenized is first isolated by PCR using primers that bear restriction sites at their 5′ termini. A second set of overlapping primers is then synthesized which is complementary to the position where the mutation is to be inserted (the changed sequence is represented by a filled circle). This gives two products, a 5′- and a 3′-derived fragment, both containing the mutation. These are then mixed, denatured and then re-annealed to give a mainly single-stranded copy of the mutant version of the gene. This is finally converted into a double-stranded form by PCR using the original two oligonucleotides as primers. This product is then cloned into a suitable vector and transformed into *E. coli*.

5.3 Phage vectors

5.3.1 M13 vectors

When the only specialized requirement is that ssDNA be easily prepared, derivatives of the bacteriophage M13 rather than the pBluescript plasmid are sometimes used as cloning vectors (*Figure 5.6*). This approach is most often used in large-scale sequencing projects. It removes the requirement for co-infection with a helper phage, but transformation efficiencies for M13 are relatively low, and large inserts are subject to frequent rearrangements during propagation. The gene of interest will usually therefore have been isolated by cloning into a vector that accepts large inserts readily. Randomly sheared DNA sub-fragments are then subcloned into M13 and ssDNA are prepared for sequence analysis.

5.3.2 In vitro *packaging of bacteriophage* λ *DNA*

The double-stranded DNA virus, bacteriophage λ, is extremely well characterized genetically and biochemically, allowing elegant methods to

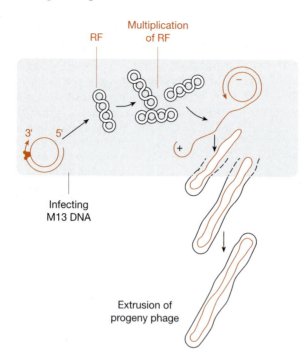

Figure 5.6. Production of single-stranded DNA. Infecting ssDNA is converted into double-stranded replicative form (RF) DNA. After replication, one strand (the + strand) is packaged into new progeny viral particles. Redrawn from Williams and Patient (1989) with permission from Oxford University Press.

be used to simplify the task of gene cloning. One of its most useful features is the ability to assemble phage particles *in vitro*. The genome is a linear, dsDNA molecule 45 kb in length. It replicates to form a long concatemer, containing many viral genomes attached end-to-end. Because the termini of the viral genome have staggered ends, they are mutually cohesive. Hence they are called *cos* sites. During packaging of the viral DNA into phage heads, the *cos* sites act as targets for cleavage of the multimeric DNA, to yield separate viral genomes. An extract prepared from infected cells will incorporate added DNA into phage coat proteins to yield viral particles. This is called *in vitro* packaging [3]. The only viral sequences required for packaging are the *cos* sites, so that vectors containing target sequences (recombinant phage) can be packaged. Using such extracts, a very high proportion of input DNA is packaged, and the introduction of the resultant phage into cells by infection (transfection) is also a very efficient process. When a complex DNA, such as the entire human genome, is the target, many hundreds of thousands of DNA segments must be cloned to have a reasonable chance of isolating a specific gene. A mixture of clones derived from a complex target is termed a clone bank, or sometimes a gene library. The high cloning efficiencies which can be obtained by *in vitro* packaging minimize the amount of DNA needed to generate such a library.

There are two kinds of vector that differ in the size of insert they accept – replacement vectors and insertion vectors. Lambda replacement vectors are not now commonly used, because they have been replaced by vectors that accept much larger inserts, but derivatives of the original insertion vectors are still in use. Both forms of vector will, however, be described, because they illustrate several important fundamental principles.

5.3.3 Lambda replacement vectors

Lambda replacement vectors contain restriction sites flanking a region of the genome which is dispensable for phage propagation in suitable bacterial host strains (*Figure 5.7*). Cloning into a replacement vector is performed in a somewhat similar manner to plasmid cloning. The phage DNA is circularized by ligation of the *cos* sites. The central, non-essential fragment is cleaved out and the linked flanking fragments (arms) are purified by centrifugation or electrophoresis. Ligation is performed at a ratio of arms to target that favors the formation of very long concatemers, in which vector and target molecules are interspersed. Because *in vitro*

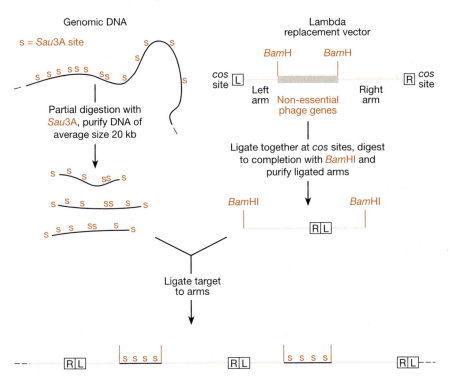

Figure 5.7. Gene cloning in a λ replacement vector. For successful genomic cloning, the target DNA must be isolated in as large a form as possible. Partial cleavage with *Sau*3A is performed by limiting either the time of digestion or amount of enzyme. In order to obtain efficient ligation and packaging, the ratio of arms to target, and their absolute concentration must be carefully controlled.

packaging requires the two *cos* sites to be a minimum of 38 kb apart, there is an automatic selection for those molecules where a target fragment is flanked at each side by a vector molecule. In most replacement vectors [4] the internal region that is replaced by the target contains a gene that renders the phage non-viable in an appropriate *E. coli* host. Hence it is possible to select against parental phage, that is, those vector molecules that do not contain target DNA, by using such a host for infection. Recombinant phage, in which the internal region is replaced by target DNA, will of course be viable.

5.3.4 Lambda insertion vectors

When cloning into an insertion vector, the phage DNA is cleaved with a restriction enzyme that cuts it only once, and the target is inserted into this site (*Figure 5.8*). Because no phage DNA is removed, and the maximal separation between *cos* sites that permits DNA to be packaged is 52 kb, there is a much lower limit on the size of target which can be cloned in insertion vectors. Hence they are often used for cloning cDNA copies of eukaryotic mRNA sequences. These have an average size of 1.5–2 kb and do not generally exceed 5 kb in length. In one typical vector, λgt10 [5], the *Eco*RI cloning site is in a gene which is deleterious to phage replication in certain host strains. This allows selection against non-recombinant phage (*Figure 5.8*).

5.3.5 A mixed lineage vector: λZAP

The λZAP vector combines the best of plasmid and phage vectors to give a highly versatile tool for the molecular geneticist. It comprises a λ phage

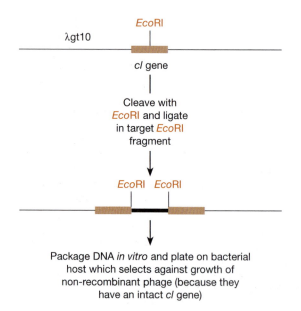

Figure 5.8. Gene cloning in a λ insertion vector. Phage containing an intact *cI* gene will not grow on the *E. coli* strain. Thus, by infecting such a host with a phage–target ligation mix, which always contains a mixture of recombinant (insert-containing) and non-recombinant (parental) phage, it is possible to filter out the unwanted non-recombinant phage completely.

containing one copy of the pBluescript molecule. Target DNA is ligated into the MCS contained within the pBluescript portion of the λZAP molecule. The ligation mix is then subjected to *in vitro* packaging. Because of the high intrinsic efficiency of the packaging reaction, a very large number of plaques are produced upon infection of *E. coli*. In addition to this high efficiency of cloning, λ phage are very suitable for some forms of screening (e.g. *in situ* hybridization). However, λ phage are not as convenient as plasmids for preparing and analyzing the target DNA molecules. Hence, once the desired plaque has been identified, the bacteriophage is introduced into cells in a co-infection with a modified M13 'helper phage'. The helper phage proteins direct the excision of the pBluescript region of the recombinant λ phage. This pBluescript:target recombinant molecule is amplified up to a high copy number and can readily be isolated from the cell in large amounts (*Figure 5.9*).

5.3.6 Cosmid, PAC and BAC vectors

Lambda replacement vectors permit cloning of DNA fragments of up to 20 kb, but very often it is necessary to isolate fragments larger than this.

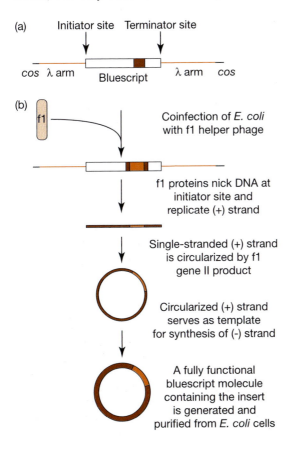

(a) Initiator site   Terminator site

cos   λ arm   Bluescript   λ arm   cos

(b) f1

Coinfection of *E. coli* with f1 helper phage

f1 proteins nick DNA at initiator site and replicate (+) strand

Single-stranded (+) strand is circularized by f1 gene II product

Circularized (+) strand serves as template for synthesis of (-) strand

A fully functional bluescript molecule containing the insert is generated and purified from *E. coli* cells

Figure 5.9. (a) The structure of the cloning vector λZAP. (b) Mechanisms of *in vivo* excision of the recombinant pBluescript vector from the λZAP molecule. *E. coli* cells are co-infected with recombinant λZAP, containing the cDNA of interest, and f1 helper phage. The f1 protein encoded by the helper genome recognizes the initiator site on the λZAP DNA and starts replication, ending at the terminator site. The single-stranded product is then circularized by gene II protein and used as template for the synthesis of the complementary strand to generate a fully functional recombinant pBluescript that can be isolated from the infected cells.

Possible reasons include: (a) many genes in higher eukaryotes are longer than 20 kb and it is highly desirable to keep them in one piece during cloning; (b) some of the procedures used to screen genomic clone banks yield clones which are close to, but not at, the position of the required locus. It is then necessary to 'walk' along the genome, by isolating a series of genomic clones which overlap, and therefore link, the starting clone and the required gene (*Figure 5.10*). It is obviously more convenient if individual steps in such a walk be as large as possible, so the genomic clones must be as long as possible.

Cosmid vectors were the first vectors used to prepare genomic DNA libraries for these purposes and they are still in current use. As their name suggests, cosmids are plasmid vectors containing the *cos* sites of bacteriophage. Target DNA fragments are cloned into cosmid vectors in much the same way as into a replacement vector, and the DNA is packaged *in vitro* and introduced into *E. coli* by transfection. The cosmid vector contains a drug-resistance gene, and recombinant clones are selected and propagated in just the same way as bacteria transformed with a plasmid vector. Again, this is a very efficient cloning method and representative libraries containing inserts of up to 50 kb in length can readily be generated [6].

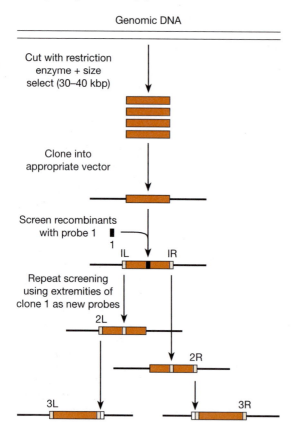

Figure 5.10.
Chromosome walking. After constructing the genomic library, the screening is performed starting from a known position (probe 1). The sequences taken from the extremities of each positive clone isolated with probe 1 generate two families of probes (L = left; R = right) with which it is possible to 'walk' along a chromosome in both directions.

In order to increase the size of insert that can be cloned beyond the approximately 50 kb limit imposed by the bacteriophage λ packaging constraint, several methods have been devised. The bacteriophage P1 packages DNA molecules of up to about 100 kb in length and P1-derived artificial chromosomes (PAC) cloning vectors (*Figure 5.11*) are sometimes used in genome cloning and sequencing projects [9]. This upper limit of 100 kb is again set by packaging constraints. Fortunately, however, methods of generating electroporation-competent bacteria have been improved to the point where large, representative libraries can be prepared without the need to package DNA. Using bacterial artificial chromosome (BAC) vectors (*Figure 5.12*) inserts of up to about 300 kb in length can be accommodated. This size of insert represents a sizeable proportion of the *E. coli* host genome itself and obviously only a few such molecules can be maintained per cell. This constraint is built into the BAC vector because it contains appropriate replication and copy number control sequences (*Figure 5.12*). This low

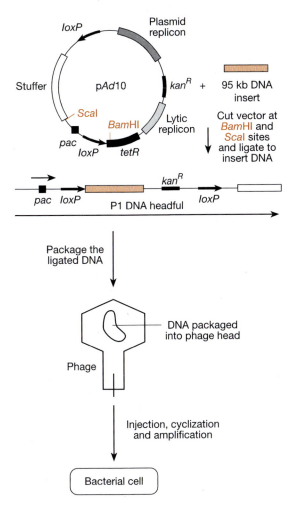

Figure 5.11. Structure of a P1 cloning vector. The vector contains a number of genes essential for propagation of the vector and for antibiotic selection. Up to 100 kb of target DNA can be inserted between the *Bam*HI and *Sca*I sites, which generates two vector arms and disrupts the tetracycline resistance gene (*tetR*). The recombinant clones are packaged *in vitro*, and used to transform bacteria to kanamycin resistance. Modified from Sternberg (1992) with permission from Elsevier.

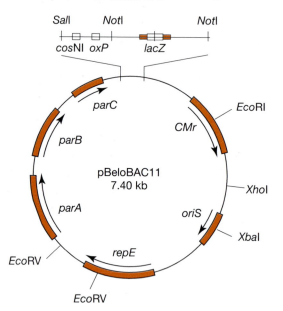

Figure 5.12. Structure of a BAC cloning vector. pBeloBAC11 is a plasmid vector based upon the F factor of *E. coli*, the naturally occuring mini-chromosome that is reponsible for genetic interchange [8]. The available cloning sites are: *Bam*HI, *Sph*I, and *Hin*dIII (B/S/H) and they are flanked by the T7 and SP6 RNA polymerase promoters. Due to its large size, the cloned insert will contain very many restriction sites. Therefore, to make it possible to cut at a single position outside the cloned DNA, the *cosN* and *loxP* sites, recognized and cut by the λ terminase and P1 Cre protein respectively, are positioned at one side of the MCS. The vector contains the *lacZ* gene for recombinant selection and *CMr* (chloramphenicol resistance) for transformant selection. Collectively, the F factor genes *oriS, repE, parA* and *parB* regulate copy number at a level of one or two per cell.

copy number is sometimes inconvenient but has one very significant compensating advantage. Homologous recombination between recombinant DNA sequences present at high copy number within a bacterial cell is a very frequent event and can lead to unwanted DNA rearrangements (see Section 5.4.3). The recombination frequency increases with the size of the cloned insert, hence the low copy number of the BAC vectors is a very important safeguard when cloning very large fragments of genomic DNA.

5.4 Construction and screening of genomic libraries

The construction and screening of genomic DNA libraries are now fairly routine procedures and it is also possible to buy libraries 'off the shelf' or to have them constructed to order using the favored DNA as the target. However, it is important to understand how this is achieved, so that the limitations inherent in using such libraries become apparent.

5.4.1 Generating target DNA for genomic cloning

The genomic DNA used as the source of the target must be carefully isolated, so that it is as large as is practically possible. Prior to cloning, it must be cleaved into pieces of a size suitable for cloning in the vector of choice. The aim is to generate hundreds of thousands of different DNA fragments that form a set of overlapping cleavage products. Every gene will then be represented within a set of fragments, most of which will have different end points (*Figure 5.7*). It is very important that the cleavage of the target DNA be at least semi-random, otherwise some genes would not be included within DNA fragments of a clonable size. This kind of 'semi-random' cleavage is often achieved by partial cleavage with the enzyme *Sau*3A (*Figure 5.7*). In DNA with a normal base composition the average length of fragments generated using *Sau*3A is 256 nucleotides. It is therefore very unlikely that a required gene will not have *Sau*3A sites spaced around it in such a way as to generate a fragment within the required size range for cloning. In practice, a controlled partial digestion with *Sau*3A is performed. Fragments of a size range compatible with the vector system to be used are purified by gradient centrifugation or gel electrophoresis.

5.4.2 Insertion of target DNA into the cloning vector and storage of the genomic library

In the case of λ phage libraries the target is ligated into the purified arms of the cloning vector (*Figure 5.7*). After ligation, the DNA is subjected to *in vitro* packaging and used to transfect *E. coli*. In the case of a cosmid library, the drug-resistant bacterial colonies are stored by replica plating the bacteria on to filter paper. This process is often termed 'lifting a library'. One copy of the library is stored frozen, so as to preserve bacterial viability. Bacteria on another, identical, filter paper copy are lysed in such a way as to cause the cosmid DNA released from each colony to become denatured and bind to the filter without diffusing away (*Figure 5.13*). The DNA on this latter filter is subjected to *in situ* hybridization, using a radioactive probe specific for the gene of interest. Once the area on the filter containing the required clone is identified, viable bacteria are isolated from the stored copy.

In the case of λ libraries, which consist of a large number of plaques where the bacteria have been lysed rather than a collection of viable bacterial colonies, the above procedure is not practicable. For long-term storage, phage particles are eluted from the plates in a mixed pool, containing representatives of every different recombinant phage on the plate. Such pooled, or amplified, libraries are very convenient since they can readily be exchanged between laboratories or sold commercially. They can then be replated onto fresh host cells, lifted onto filters and screened. However, phage plaques containing different recombinant DNA molecules are almost invariably of different size, because individual cloned

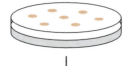

Agar plate with bacterial
colonies or phage plaques

|

Transfer the bacterial colonies or phage
plaques to filter and lyse to release DNA

|

Hybridize with labeled probe 'a'

|

Perform autoradiography
with labeled filter

Phage plaque or
bacterial colony
containing nucleic
acid 'a'

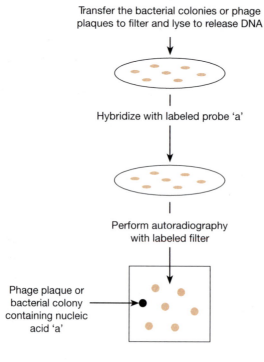

Figure 5.13. Screening clones by *in situ* hybridization. This technique is equally applicable to phage plaques or bacterial colonies. When screening a library, the density of plaques or phage is very much higher than shown here, with perhaps 100 000 colonies or plaques on a 15-cm diameter plate. After screening, a plug of agar is lifted out and the bacteria or phage eluted and replated at lower density. This process is repeated until a pure bacterial clone or phage population is obtained.

segments of genomic DNA may affect the extent of phage replication differently. Hence, amplified libraries will be biased in favor of some sequences and against others. The extent of the bias obviously increases with the number of rounds of amplification a library has undergone. It is therefore quite dangerous to use such a library to isolate a genomic clone. The sequence required may be seriously under-represented or even absent from the library. An unamplified library (e.g. a newly packaged recombinant phage), or a filter copy of a cosmid library, are much safer, if sometimes less convenient, alternatives.

5.4.3 Unclonable DNA and DNA rearrangements in E. coli

The problem of bias in genomic clone banks is part of a larger issue: how reliable is *E. coli* as a cloning host? There are sequences in mammalian DNA that are unclonable in *E. coli* and this represents a major potential problem in genomic cloning. Despite its importance, this is still an area where there are many uncertainties. Obviously, if a piece of DNA cannot be isolated by gene cloning it is very difficult to determine the nature of the problem. There are, however, general prejudices and some hard facts which can perhaps give an insight.

If a segment of target DNA encodes a protein that is lethal to *E. coli*, and if this gene happens to lie downstream of vector sequences that can direct transcription and translation in *E. coli*, then the recombinant may never be recovered. This problem will obviously be worsened if the vector is present at high copy number, because there is normally a good proportionality between the number of copies of the gene in a cell and the amount of its protein product. Copy number is determined by plasmid-encoded sequences, for example pBluescript is normally present at several thousand copies per cell. There are, however, vectors that replicate to give many fewer copies per cell [8] and these are sometimes used to circumvent this problem.

Gene toxicity will sometimes lead to the absence of a particular recombinant in a gene bank, but most frequently causes the host bacterium to grow relatively slowly. This is another major reason why banks should always be screened prior to amplification if at all possible. An additional reason for avoiding amplification is the potential instability of DNA during propagation in the host. This is a particularly acute problem in the case of DNA that contains repetitive sequences. If, for example, two highly homologous genes are closely linked on the same piece of DNA in the same relative orientation, then they may undergo recombination to yield a single 'hybrid' gene. This will result in the introduction of a deletion into the cloned sequence, spanning the 3' end of one gene, the intervening DNA and the 5' end of the other gene (*Figure 5.14*). If the two genes are in the opposite relative orientation, then the host cell machinery may also catalyze recombination between the two genes, but in this case the substrate for recombination is the hairpin structure that can be formed by the two homologous sequences (*Figure 5.14*).

It is relatively rare for two highly homologous genes to be situated close together in the genome and, in order to be substrates for homologous recombination, two sequences must be identical along a considerable portion of their length. However, if a genome contains many copies of a simple repeat sequence, it is frequently subject to rearrangement in *E. coli* because the repeats provide regions of high homology. Recombinational deletion may occur at any time during the propagation of a particular DNA in a particular host, so that by the time a recombinant DNA is first analyzed it may already contain a deletion. It is therefore sometimes desirable to go back to the genomic DNA from which the gene was cloned and analyze the structure of the authentic gene by performing Southern blotting. If the cloned DNA and the genomic DNA share the same restriction map then it is reasonably safe to assume that no gross deletion has occurred.

This may all seem like a litany of potential disaster but the risk of recombinational rearrangement can be minimized by using *E. coli* mutants

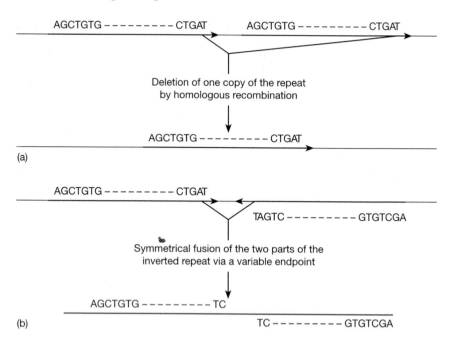

Figure 5.14. Generation of rearrangements by homologous recombination.
(a) Direct repeats. If two pieces of DNA lie in the same relative orientation, and the DNA loops back on itself such that the two homologous pieces of DNA lie side by side, then the *E. coli* recombination system will recognize this as a substrate for homologous recombination. The result will be to excise one copy of the repeat and the intervening DNA. (b) Inverted repeats: if two pieces of DNA lie in the opposite relative orientation and the DNA becomes supercoiled (see Chapter 1) then DNA may loop out and rearrange to form a cruciform, or snapback, structure. Such structures are also substrates for the recombination system, but the extent of the deletion may vary between clones, depending upon the site in the cruciform structure at which the rearrangement occurs.

with lesions in the genes catalyzing recombination. Some *E. coli* strains contain mutated versions of several different genes required for homologous recombination. If such a strain is used as a host, then the likelihood of rearrangement is greatly reduced. A price has to be paid for this in that these mutants grow relatively poorly, but this price is well worth paying if cloning artifacts can thus be avoided.

References

1. **Vieira, J. and Messing J.** (1982) The pUC plasmids, an M13mp7-derived system for insertion mutagenesis and sequencing with synthetic universal primers. *Gene* **19**: 259.
2. **Shimada, S.** (1996) PCR-based site-directed mutagenesis *Methods Mol. Biol.* **57**: 157–165.
3. **Hohn, B. and Murray, K.** (1977) Packaging recombinant DNA molecules into bacteriophage particles *in vitro. Proc. Natl Acad. Sci. USA* **74**: 3529.

4. **Frischauf, A.M., Lehrach, H., Poutska, A. and Murray, N.** (1983) Lambda replacement vectors carrying polylinker sequences. *J. Mol. Biol.* **170**: 827.

5. **Huynh, T.V., Young, R.A. and Davis, R.W.** (1985) *DNA Cloning: a Practical Approach*, Vol. 1 (ed. D.M. Glover). IRL Press, Oxford.

6. **Ish-Horowicz, D. and Burke, J.F.** (1981) Rapid and efficient cosmid cloning. *Nucleic Acids Res.* **9**: 2489.

7. **Burke, D.T., Carle G.F. and Olson M.V.** (1987) Cloning of large segments of exogenous DNA into yeast by means of artificial chromosome vectors. *Science* **236**: 805–811.

8. **Shizuya, H., Birren, B., Kim, U-J., Mancino, V., Slepak, T., Tachiiri, Y. and Simon, M.** (1992) Cloning and stable maintenance of 300-kilobase-pair fragments of human DNA in *Escherichia coli* using an F-factor-based vector. *Proc. Natl Acad. Sci. USA* **89**: 8794–8797.

9. **Ioannou, P.A., Amemiya, C.T., Granes, J., Kroisel, P.M., Shyzuya, H., Chen, C., Batzer, M.A. and de Jong, P.J.** (1994) A new bacteriophage P1-derived vector for the propagation of large human DNA fragments. *Nature Genet.* **6**: 84–89.

10. **Vos, J.M.** (1998) Mammalian artificial chromosomes as tools for gene therapy. *Curr. Opin. Genet. Devel* **8**: 351–359.

Further reading

Berger, S.L. and Kimmel, A.R. (1987) *A Guide to Molecular Cloning Techniques. Methods in Enzymology*, Vol. 152. Academic Press, San Diego, CA.

Miller, J.H. (1972) *Experiments in Molecular Genetics*. Cold Spring Harbor Laboratory Press, New York.

Sambrook, J., Fritsch, E.F. and Maniatis, T. (1989) *Molecular Cloning, a Laboratory Manual*. Cold Spring Harbor Laboratory Press, New York.

Winnacker, E.L. (1987) *From Genes to Clones: Introduction to Gene Technology*. VCH, Weinheim.

Chapter 6

Cloning in *E. coli* II: isolating genes

6.1 Construction of cDNA libraries

If total human genomic DNA is digested to completion with a restriction enzyme that cuts it into pieces averaging 40 kb in size, over 100 000 different fragments are generated. Cloning would allow each of these to be isolated separately, but screening them individually for a specific gene would be a gargantuan task. Also, eukaryotic genes are generally interrupted by introns, and bacteria cannot splice eukaryotic gene transcripts. This precludes a very powerful screening method based upon gene expression in bacteria. Genomic cloning is not, therefore, used as the initial method of identifying and isolating eukaryotic genes. They are usually isolated by cloning complementary DNA copies (cDNA clones) of their cognate mRNA sequences. The cDNA clone is then used as a hybridization probe to screen a genomic library for the required gene.

6.1.1 Purification of mRNA for cDNA cloning

In order to minimize the numbers of clones which must be prepared and screened, mRNA is normally purified from a type of cell in which the required sequence is maximally abundant. This flexibility of choice constitutes one of the great advantages of cDNA cloning over genomic cloning. In some specialized tissues one, or just a few, mRNA sequences will be highly abundant. In a reticulocyte, for example, the α- and β-globin mRNAs constitute the vast bulk of the mRNA population.

As with the preparation of genomic libraries, it is very important that the mRNA be isolated in as intact a form as possible. The great enemies here are ribonucleases. These are released when cells are lysed with detergents, or they can be introduced through contamination during mRNA purification. There are, however, potent RNAase inhibitors and well-established practices for circumventing both these problems. The mRNA is purified away from proteins and lipids using organic solvents or by centrifugation in the presence of powerful denaturants. The mRNA constitutes only about 1–2% of total cellular RNA, so it is often purified further, away from rRNA and other structural RNAs, by affinity chromatography on short tracts of dT homopolymer (oligo-dT) bound to an inert support. The mRNA anneals to the oligo-dT at high salt concentration, by virtue of its poly(A) tail, while rRNA and tRNA are not bound. The mRNA is then eluted in a low salt buffer.

6.1.2 Synthesis of double-stranded cDNA

Affinity purification of mRNA is not an essential step in cDNA cloning, because the poly(A) tract can be used to ensure that the mRNA is selectively copied into cDNA (*Figure 6.1*). An excess of oligo-dT is added as primer and this anneals to the poly(A) tail. The enzyme reverse transcriptase is used to copy this primer–template complex, to yield a cDNA copy representative of the entire mRNA population. There are a number of available procedures whereby cDNA may be used to generate double-stranded DNA suitable for cloning. One commonly used method for preparing double-stranded cDNA is shown in *Figure 6.1*. The mRNA and cDNA remain annealed after the reverse transcriptase reaction, and the enzyme RNAaseH is added to introduce breaks into the RNA. The

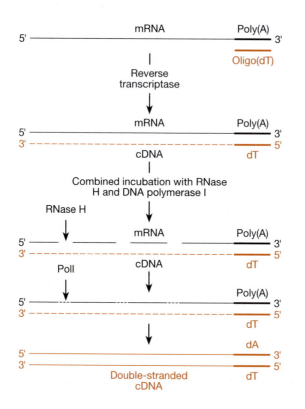

Figure 6.1. The preparation of double-stranded cDNA. Redrawn from Williams and Patient (1989), with permission from Oxford University Press.

resulting nicked molecules act as primer–template complexes for the enzyme DNA polymerase-1, and this produces a double-stranded cDNA copy of the mRNA.

In order to provide cohesive termini, and hence increase the efficiency of ligation into the vector, chemically synthesized restriction cleavage sites (linkers) are often ligated onto the cDNA (*Figure 6.2*). The restriction enzyme site present within the linker will sometimes occur fortuitously within some cDNAs. Since it is obviously desirable to clone an intact copy of the mRNA, the double-stranded cDNA is treated with a restriction enzyme methylase before addition of the linkers. In the example shown in

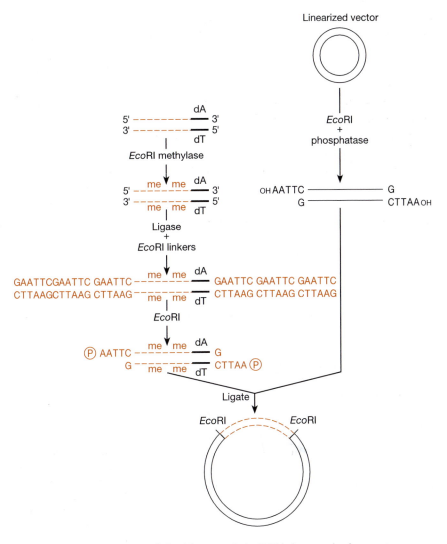

Figure 6.2. The insertion of double-stranded cDNA into a cloning vector. Redrawn from Williams and Patient (1989) with permission from Oxford University Press.

Figure 6.2, where *Eco*RI linkers are used for cloning, the target is modified with *Eco*RI methylase. This introduces methyl groups into every *Eco*RI recognition site within the cDNA. After ligation to the methylated target, the linkers are cleaved with *Eco*R1, while the methylated cDNA remains uncleaved. This is just one of many possible routes to efficient cDNA insertion and there are commercially available kits for all these methods. Some of these methods are particularly powerful because they incorporate a PCR step and this means that tiny amounts of mRNA can be used as templates for the cDNA preparation.

6.1.3 Insertion of the double-stranded cDNA into the vector

The double-stranded cDNA is ligated into the vector of choice, commonly into a λ insertion vector such as λZAP. This readily accepts cDNA-sized fragments and, because it can be introduced into bacteria by *in vitro* packaging, has a very high efficiency of cloning. This is important if only limited amounts of mRNA are available for preparing the library. In the case of cDNA libraries there is generally less of a problem with selective loss or under-representation if the library is pooled. Hence it is reasonably safe to use amplified libraries, provided the initial complexity (that is, the number of total clones which were pooled) is high and not too many cycles of amplification are used.

6.2 Screening cDNA libraries

The preparation of the library is most often the least problematical of the steps in isolating a particular cDNA clone (and the problem of course vanishes if a reliable amplified library is available). In contrast, screening a library to identify the required sequence is often very difficult and time-consuming. The scale of the task, and the nature of the strategy used, depend on two factors: the abundance of the required sequence in the cell type used to isolate the mRNA for cloning, and the level to which the cognate protein has been characterized. The following strategies are commonly employed.

6.2.1 Differential screening

The cloning strategy which necessitates the least prior information about the sequence to be cloned is differential screening [1]. Its only requirement is that two cell types are available which differ in their level of expression of the desired sequence. The two cell types might be, for example, a hormone-responsive cell line which is inducible for a particular mRNA and the same cell line cultured in the absence of hormone. First, a cDNA library is prepared from hormone-induced cells and duplicate lifts of the library are made (*Figure 6.3*). One copy filter is screened with radioactively

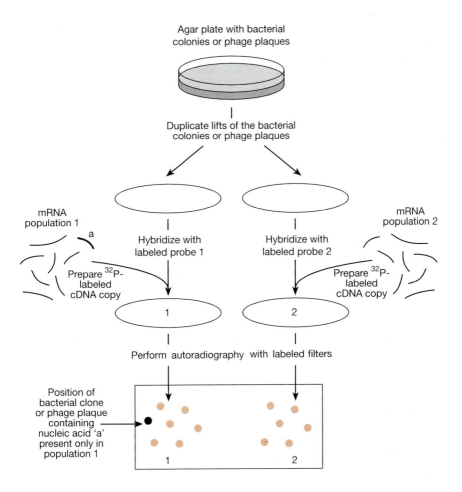

Figure 6.3. Differential screening of cDNA libraries. Two copies of the recombinant population are made by lifting onto a hybridization membrane. The filters are separately screened with a labeled probe made from each of the two test RNA populations. The resulting autoradiograms are overlaid to search for signals which are present on only one of the two filters. The clone or plaque responsible for this signal is then isolated by eluting the phage or colonies, replating at lower density (for clarity this diagram shows only a few recombinants; in practice a plate might bear 10 000 recombinants) and rescreening.

labeled cDNA prepared using mRNA isolated from the induced cells, and the other is screened with labeled cDNA prepared from the uninduced cells. A cDNA clone derived from an mRNA which accumulates only in the presence of the hormone will show a hybridization signal with the probe prepared from the induced cell mRNA, but no signal with the probe prepared from uninduced mRNA. Differential screening can also be used to identify sequences which show quantitative, rather than qualitative, variation between two cell types. The most abundant mRNA sequences direct synthesis of the most cDNA. Therefore the amount of cDNA which hybridizes to a colony is determined by the abundance of that particular sequence in the starting mRNA population. An autoradiogram obtained in differential screening shows signals which vary in strength over several orders of magnitude, reflecting differences in the abundance of individual sequences in the starting mRNA population. By comparing autoradiograms, it is possible to detect differences of as little as three to five-fold in signal strength between duplicate lifts. This is therefore a very useful method of analysis. It allows the patterns of gene expression in two cell types to be compared, and allows genes which show even a small differential expression to be isolated.

A variant of the differential screening technique is to remove sequences which are common to the two mRNA populations. A number of strategies have been used for this. In the case described above, the simplest method would be to hybridize an excess of the uninduced cell mRNA to induced cell cDNA (*Figure 6.4*). Sequences which hybridize can then be removed by, for example, column chromatography on hydroxyapatite (a matrix which allows the separation of single-stranded nucleic acids away from double-stranded molecules). The unhybridized, single-stranded cDNA is then used as a substrate for cDNA cloning. The resulting library, usually called a subtraction library, can be analyzed by differential screening using subtracted cDNA and control, unsubtracted cDNA. The great advantage of this method over differential screening is that far fewer clones need to be screened. It is a particularly useful technique for obtaining an enriched source of mRNA for sequences expressed at only a few copies in each cell. However, it is applicable only when there is a qualitative difference, or a very large quantitative difference, in the abundance of the required sequence in the two mRNA populations (otherwise the required sequence would be lost during the subtractive hybridization).

Subtraction hybridization can also be used to isolate gene products characteristic of a particular differentiated cell type, provided that a closely matched, non-expressing cell type is available. A good example of this approach is provided by the cloning of the T-cell receptor [2]. T-cell receptors are expressed on the surface of mature T lymphocytes and not on B lymphocytes. Both cell types are otherwise very similar and express many of the same gene products. cDNA was prepared from T-lymphocyte mRNA and hybridized with mRNA isolated from B lymphocytes (*Figure 6.4*). The unhybridized, single-stranded cDNA was isolated by hydroxyapatite chromatography and used to construct a cDNA library. Clones for the T-cell receptor were then isolated by screening this library with a cDNA probe prepared from the original T lymphocytes.

There is a third method of isolating the DNA copies of mRNA sequences that are selectively enriched in one of two mRNA populations, called 'differential display' (*Figure 6.5*). It is based upon selective

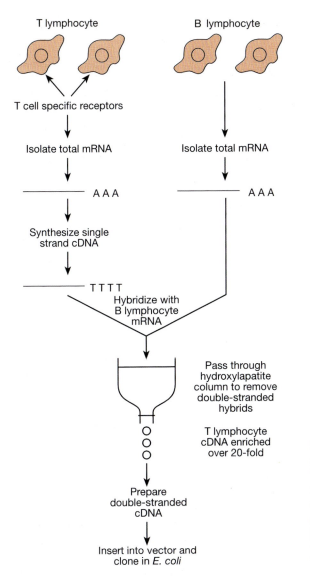

Figure 6.4. Isolation of the T-cell receptor. This shows the strategy used to obtain an enriched source of cDNA from a T lymphocyte to allow the isolation of the T-cell receptor.

amplification by PCR of the 3′ ends of mRNAs and resolution of the PCR products on an acrylamide gel. In the first step, reverse transcription, the mRNA population is copied using four different primers. These are designed to initiate transcription from the <u>same</u> relative position within <u>different</u> mRNA templates. The four primers each contain a tract of dT that is sufficiently long to prime the reverse transcriptase reaction (e.g. of 12 nucleotides in length). The four primer sets differ at their 3′ ends, where they contain a dinucleotide of variable composition. In each set the dT tract is followed by a 'random' nucleotide, N (generated by incorporating equal amounts of all four nucleotides at that particular point in the

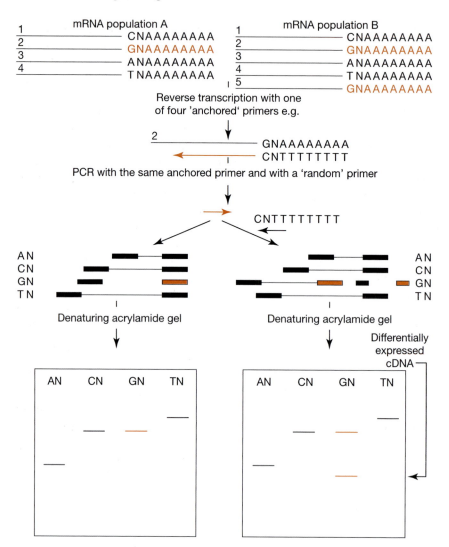

Figure 6.5. Comparison of two mRNA populations by differential display. Anchored primers are used to select specific mRNA 3′ ends from within a population of mRNAs. The product of reverse transcription is then amplified with the same anchored primer, using random sequence oligonucleotides as the upstream primers. Two populations of mRNAs A and B are compared. The primer $T_{12}NG$ detects one mRNA in population A and two mRNAs in population B. When run on acrylamide gel, the two resulting cDNA populations display a different pattern of bands in lane GN, where there is an additional band (arrowed) corresponding to the additional mRNA present in population B.

chemical synthesis reaction), and the 3′ terminal residue is either a dA, a dG, a dC or a dT residue. Using such primers the oligo-dT tract hybridizes at a fixed position on the poly(A) tails but each primer hybridizes to only a very small sub-set of the mRNA molecules. This is determined by the two residues at the end of the 3′ non-coding region of the mRNA. These must be the complement of the two 'anchor' residues at the 3′ end of the primer. Each of the four primer sets (T12NA, T12NT, T12NG and T12NC) will therefore identify a different sub-set of specific partners within the total mRNA population.

Four separate reverse transcription reactions are run for each RNA sample and the resulting cDNAs are subjected to PCR. One of the primers used in the PCR reaction is the primer used in the initial cDNA synthesis. The second primer is a mixed population, consisting of 12–16-mers in which the 6–8 bases at the 3′ end are equal mixtures of all four nucleotides at every position (i.e. N^{6-8}) and the 6–8 bases at the 5′ end are identical in all the molecules. The latter sequence sometimes incorporates a restriction site useful in the subsequent cloning steps. Given the frequency of a random 7-mer, each possible pair of primers will selectively amplify 50–100 different mRNAs. The resulting amplified cDNAs are radioactively labeled during their synthesis and resolved on a polyacrylamide sequencing gel. The pattern of bands displayed by each primer combination is specific for any given RNA population. The cDNAs of interest, that is, the bands that show a difference in the two mRNA samples being compared, are recovered from the gel and amplified by PCR. They can then be sequenced directly or cloned into a suitable vector.

Differential screening, subtraction hybridization and differential display are very useful techniques for identifying any gene with a differing level of expression in two cell types and for producing a preliminary enrichment for a particular sequence. However, their utility as a means of isolating a specific gene is critically dependent upon the two populations being very similar (aside of course from the sequence or sequences of interest). Two closely related populations are not always available and so other, more generally applicable, strategies have been devised. There are two common approaches, oligonucleotide hybridization and expression screening.

6.2.2 Screening cDNA libraries by oligonucleotide hybridization

Very powerful techniques are available for determining the amino acid sequences of proteins. If some or all of the sequence of the protein of interest can be determined, an oligonucleotide can then be synthesized chemically which contains the predicted cognate mRNA sequence. This oligonucleotide can then be used as a hybridization probe to screen a cDNA library (see, for example [3]). This sounds wonderfully straightforward but, in practice, there is a large element of luck inherent in the method. The amount of work involved depends greatly upon the nature of the known protein sequence. Most amino acids can be encoded by more than one codon (Chapter 1), so that the code is ambiguous when read in the protein to mRNA direction. Leucine, for example, can be encoded by no fewer than six different codons (*Table 1.3*). If one is fortunate, the protein sequence established will include a short region containing several amino acids encoded by minimally redundant codons, particularly methionine (AUG) or tryptophan (UGG). Even when such codons are present, it is normally necessary to synthesize mixed oligonucleotides containing more than one nucleotide at positions within the chain where codon usage is ambiguous (*Figure 6.6*). The oligonucleotide is hybridized to lifts of the cDNA library under conditions which minimize the amount of non-specific annealing to other genes. However, some non-specific hybridization will normally occur, and many cDNA clones may have to be subjected to more detailed examination before the required clone is identified.

6.2.3 Expression screening of cDNA libraries

This is a very powerful approach to cDNA cloning when a high-titer antibody is available which is specific for the required protein. A library is constructed in a vector which directs a high level of expression of the proteins encoded by cDNA inserts cloned into it. Most often this is achieved by cloning the cDNA as part of a *lacZ* fusion protein in an expression vector such as λgt11 [4]. The cDNA is cloned into the *lacZ* gene in λgt11 to yield a hybrid protein, containing the region of the *lacZ* gene which encodes the N-terminus of β-gal fused to the eukaryotic cDNA. The *lacZ* gene provides the signals required for efficient transcription and translation in *E. coli*. After infection by the recombinant phage, cells are exposed to a galactoside which induces expression from the *lacZ* promoter (Chapter 1). The fusion proteins, released from the cell by phage lysis, are lifted on to a filter and detected by binding of the specific antibody (*Figure 6.7*). Plaques which show a positive signal are then purified and recombinant DNA extracted. The final definitive proof that the positive clone is the right one is to show that the cDNA insert has the expected nucleotide sequence. Hence, it is normally necessary to know a part of the amino acid sequence of the required protein.

The series of steps leading from purification of the mRNA, first to the isolation of a cDNA clone and then to a genomic clone, is summarized in *Figure 6.8*. This is given as a guide, but almost every cloning project will incorporate variations, and whole avenues of approach have not been considered. PCR technology can be applied to generate libraries from tiny

Met Gln Tyr Glu Trp Cys Gln

5' ATG CA$_A^G$ TA$_C^T$ GA$_A^G$ TG TG$_C^T$ CA$_A^G$3'

Figure 6.6. A multiply redundant synthetic oligonucleotide for use in oligonucleotide screening. Redrawn from Williams and Patient (1989) with permission from Oxford University Press. More detailed examination (such as sequence analysis of mini-preparations of DNA) is required before the required clone is identified.

Overlay phage plaques with
filter impregnated with IPTG

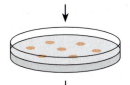

Incubate at 37°C
and remove filter

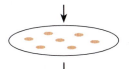

Incubate with antibody and then
with antibody detection system

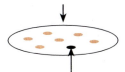

Position of positive plaque – align filter
with agar and pick plaque

Figure 6.7. Screening λ expression libraries. λgt11 contains the *lacZ* promoter so gene expression is induced with the galactoside IPTG. The induced protein, released from the lysed cells, then sticks to the filter.

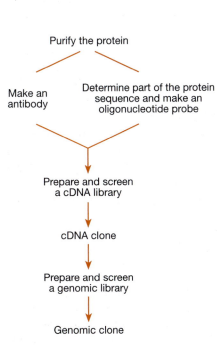

Purify the protein

Make an antibody

Determine part of the protein sequence and make an oligonucleotide probe

Prepare and screen a cDNA library

cDNA clone

Prepare and screen a genomic library

Genomic clone

Figure 6.8. Two common gene cloning strategies. These are two very commonly used gene isolation routes, but there are several alternative techniques which can be used if the protein is not easily purifiable.

amounts of starting mRNA and there are very elegant screening strategies based upon cloning cDNA in vectors which direct expression in mammalian cells [5]. The preparation and screening of cDNA libraries is a constantly evolving and vitally important part of genetic engineering technology, and there will doubtless be many further advances. However, present standard procedures make it possible virtually to guarantee isolation of any sequence, provided only that some enzymatic or biological activity can be used as the basis of purification and/or antibody production.

6.3 Identifying genes by positional cloning: cystic fibrosis

All of the gene cloning methods described thus far rely on some knowledge of the encoded protein. Minimally, this would be a knowledge of its distribution in various tissues, so that differential screening could be used. In the best possible case some, or all, of the sequence of the protein would be known. There are, however, many inherited diseases where, until the gene is cloned, there is little or no information concerning the defective protein. CF is a good example. The gene was cloned using only a knowledge of its general chromosomal location [6–8]. This provides a good example of the enormous power of this procedure of mapping followed by positional cloning.

CF affects approximately 1 in 2000 live births among Caucasians, and has a calculated carrier frequency of 5% (see also Chapter 4). The disease is associated with an abnormal function of chloride channels in apical membranes of epithelia, and electrophysiological studies have suggested the existence of an altered ion channel in CF patients. Membrane proteins are notoriously difficult to purify and, before the gene was cloned, there was no direct biochemical evidence to support this hypothesis, nor even to show that altered membrane function was the primary cause of the disease. The locus was assigned to the long arm of chromosome 7 by identifying genetic markers which tended to co-segregate with the gene in families with several affected children (*Figure 6.9*). Two markers that are closely linked in the genome will rarely be separated by recombination. By screening DNA from many families with two or more affected children

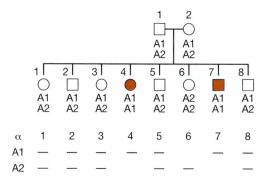

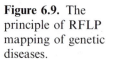

Figure 6.9. The principle of RFLP mapping of genetic diseases.

with a number of RFLP probes that are known to derive from the long arm of chromosome 7, it was possible to identify four RFLPs that showed particularly close linkage to the CF mutation. Even closely linked loci in man can still be separated by a large amount of DNA. On average, human loci which show only 1% recombination are 1 Mb apart. The linkage defined a region of about 500 kb which was expected to contain the CF locus, and this was cloned by a large-scale program of chromosome walking. It was then necessary to identify potential gene sequences within this region which might encode the CF protein. There are a number of general approaches that can be used to solve such a problem:

(i) Direct cDNA cloning – if there is a candidate tissue where the gene is expected to be expressed, a cDNA library is prepared from this tissue, and clones are isolated that hybridize with the chromosomal region known to contain the mutation.

(ii) Identification of altered mRNA transcripts in affected individuals – this is sometimes a useful approach but, again, it is only possible if there is a candidate tissue in which the gene is expected to be expressed. Detection of a difference by Northern transfer analysis relies on there being a detectable lesion, such as absence of the transcript or a size shift, in one of the avaliable mutant alleles. There are, however, more complex hybridization techniques that are capable of detecting point mutations and very small size shifts (e.g. see Section 4.2.2).

(iii) Inter-specific cross-hybridization – the coding regions of genes are conserved between closely related species, while intergenic regions and introns tend to diverge rapidly. By performing Southern transfer with DNA isolated from various mammalian species (a 'zoo blot'), using different parts of the candidate region as a probe, it is sometimes possible to identify potential coding sequences.

(iv) Identification of CpG islands – many genes have large numbers of CpG dinucleotides in the region encompassing their start sites, often because this area is rich in Sp1 sites (Section 2.4.1). Such sequences are known as CpG islands, because this dinucleotide is under-represented elsewhere in human DNA. Restriction enzymes that contain CpG in their recognition sequences preferentially cut DNA within these islands, so by screening clones with such enzymes, it is sometimes possible to identify genes.

(v) Direct identification of a potential ORF by conventional DNA sequencing – once candidate coding regions have been identified using the techniques described above, then it becomes feasible to determine their DNA sequence. This might appear straightforward but it is very difficult in practice. The sequence elements that signal initiation of transcription, gene splicing and polyadenylation are not absolutely conserved. Hence it is not possible to be certain that a predicted ORF really does form part of a gene. The only reliable methods are to perform exhaustive RNA mapping, or to isolate the cognate cDNA clone from an expressing tissue.

In the case of CF, inter-specific cross-hybridization yielded four potential candidate genes. Additional genetic evidence eliminated one of these, and another did not contain an ORF of appreciable size. The third fragment cross-hybridized weakly to other mammalian DNAs, showed a high frequency of CpG islands and the presence of several ORFs, but did not hybridize to a transcript in any of the human tissues tested. The fourth fragment contained CpG-rich regions and hybridized to a cDNA clone in libraries prepared from several tissues. When the cDNA clone was used as a probe in Northern transfers it detected a 6.5 kb transcript in many of the tissues tested. Strong evidence that this was the CF gene came from sequence analysis in affected individuals. Approximately 70–80% of CF-bearing chromosomes in northern Europeans carry a 3 bp deletion that results in elimination of a phenylalanine at position 508 of the deduced protein product. This deletion is never seen on normal chromosomes, and it is associated with the most severe form of the disease. Final confirmation that this was indeed the CF gene came from functional studies. Analysis of the predicted protein sequence suggested it was a transmembrane protein with an adenosine triphosphate (ATP)-binding site in the cytoplasmic domain (*Figure 6.10*). This structure is consistent with it being an ion-channel protein. Direct evidence for this derived from experiments where the gene was expressed in cultured cells from CF patients. A wild-type, but not a mutant, form of the cDNA complemented the defect in the chloride channel shown to be characteristic of CF. Thus a massive effort, involving several different laboratories and a combination of many skills, revealed the cause of this long mysterious disease. It also made possible the construction of specific probes for CF that allow accurate pre-natal diagnosis of mutant alleles in families with a history of the disease (Section 4.2.1).

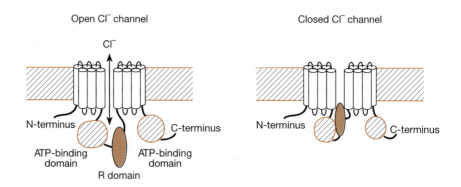

Figure 6.10. Model of CFTR function and regulation. Cystic fibrosis is caused by a defect in chloride ion secretion by epithelial cells. The CFTR gene is an integral membrane protein whose function is as an ion transporter. It contains two membrane-spanning domains and a regulatory domain (R) which controls the flow of chloride ions across the membrane in response to cAMP-activated protein kinases. Reproduced from T. Strachan (1992) *The Human Genome*, p. 132, with permission.

References

1. **Williams, J.G. and Lloyd, M.** (1979) Changes in the abundance of polyadenylated RNA during slime mould development measured using cloned molecular hybridization probes. *J. Mol. Biol.* **129**: 19.
2. **Hedrick, S.M., Cohen, D., Nielsen, E.A. and Davis, M.M.** (1984) Isolation of cDNA clones encoding T cell-specific membrane-associated proteins. *Nature* **308**: 149.
3. **Ullrich, A., Berman, C.H., Dull, T.J., Gray, A. and Lee, J.M.** (1984) *EMBO J.* **3**: 361.
4. **Huynh, T.V., Young, R.A. and Davis R.W.** (1985) *DNA Cloning, a Practical Approach* (ed. D.M. Glover). IRL Press, Oxford.
5. **Seed, B.** (1987) An LFA-3 cDNA encodes a phospholipid-linked membrane protein homologous to its receptor CD2. *Nature* **329**: 840.
6. **Rommens, J.M., Ianuzzi, M.C., Kerem, B.-S.** *et al.* (1989) Identification of the cystic fibrosis gene: chromosome walking and jumping. *Science* **245**: 1059.
7. **Riordan, J.R., Rommens, J.M., Kerem, B.-S.** *et al.* (1989) Identification of the cystic fibrosis gene: cloning and characterization of complementary DNA. *Science* **245**: 1066.
8. **Kerem, B.-S., Rommens, J.M., Buchanan, J.A.** *et al.* (1989) Identification of the cystic fibrosis gene: genetic analysis. *Science* **245**: 1073.

Further reading

Collins, F.S. (1992) Cystic fibrosis: molecular biology and therapeutic implications. *Nat. Genet.* **1**: 3.

Collins, F.S. (1992) Positional cloning: let's not call it reverse anymore. *Science* **256**: 774.

Matz, M.V. and Lukyanov, S.A. (1998) Different strategies of differential display: areas of application. *Nucleic Acids Res.* **26**(24): 5537–5543.

Sambrook, J., Fritsch, E.F. and Maniatis, T. (1989) *Molecular Cloning, a Laboratory Manual.* Cold Spring Harbor Laboratory Press, New York.

Wicking, C. and Williamson, R. (1991) From linked marker to gene. *Trends Genet.* **7**: 288.

Williams, J.G. and Patient, R.K. (1989) *Genetic Engineering.* IRL Press, Oxford.

Zhang, J.S., Duncan, E.L., Chang, A.C. and Reddel, R.R. (1998) Differential display of mRNA. *Mol. Biotechnol.* **10**(2): 155–165.

Chapter 7

Cloning in higher organisms

E. coli is the workhorse of genetic engineering but, because chromosome structure and gene expression in prokaryotes and eukaryotes differ so radically, there are some inevitable limitations to its usefulness. Hence methods of introducing and stably maintaining foreign DNA have been developed for most of the commonly studied eukaryotic organisms. As with bacterial cloning, the process is called transformation, but here the term should not be confused with the phenotypic transformation induced by viral and cellular oncogenes.

While the general principles of gene cloning are similar, the eukaryotic cell presents problems and opportunities not applicable to prokaryotes. The most obvious difference is the presence of a nuclear membrane. Fortunately this has not proven a major bar to transformation. When DNA is introduced into eukaryotic cells it readily crosses the nuclear membrane in some as yet undefined manner. Alternatively, DNA can be introduced directly into the nucleus by micro-injection. The ability to transform eukaryotes has opened up many new possibilities. Thus it is now possible to study the expression of cloned genes in their natural milieu and to modify the phenotype of the host organism.

7.1 Cloning in yeast

The budding yeast, *Saccharomyces cerevisiae*, has many attractions as a host for genetic engineering. The classical genetic approaches, which can best be applied in a haploid cell, have made it the organism of choice for investigating many fundamental processes in the eukaryotic cell. As with *E. coli*, the spin-off from this basic research has been the establishment of a battery of highly sophisticated molecular genetic techniques.

7.1.1 Gene replacement in yeast

Yeast cells grow with a doubling time of just 90 min and can be maintained in either a haploid or diploid state. They have a rigid outer cell wall, which is removed by enzymatic treatment, and the DNA is introduced by incubation in a mixture of calcium chloride and polyethylene glycol. Yeast cells can be grown in the laboratory in a simple defined medium and strains have been isolated which are incapable of growth in media lacking specific amino acids. These auxotrophic strains contain mutations in the genes responsible for specific amino acid biosynthesis; for example, a *TRP*− strain is incapable of growing in the absence of added tryptophan. The *TRP1* gene encodes an enzyme in the tryptophan biosynthesis pathway. When the *TRP1* gene is introduced into a *TRP*− cell by DNA transformation it corrects the defect, yielding cells which are rendered capable of growth in the absence of added tryptophan.

The simplest kind of yeast vector [1] contains a selectable marker, such as *TRP1*, joined to a bacterial plasmid vector (*Figure 7.1*). A hybrid, or 'shuttle', vector of this kind can be used to transform either yeast or *E. coli*. Almost all eukaryotic vectors are of this form, because it is much easier to manipulate DNA by performing genetic engineering in *E. coli*. Bacterial transformation efficiencies are very high and bacteria grow more quickly than eukaryotic cells. In a typical experiment a gene under investigation would be ligated into a *TRP1* vector and cloned in *E. coli*. It might then be subjected to further genetic manipulation, such as the introduction of specific mutations, again using *E. coli* as host. Finally, it would be transformed into yeast cells to determine the effects of the mutations on the function of the gene.

When transformed into yeast, such a vector integrates into the genome by recombination. Recombination is mediated by the battery of enzymes that naturally act to exchange DNA between homologous chromosomes. A single homologous recombinational event generates two copies of the *TRP1* gene linked in tandem (*Figure 7.1*). Transformants generated by such an event are very unstable in the absence of selection (that is, when grown in the presence of tryptophan), because recombination between the two tandemly repeated sequences occurs at a very high rate. This results in excision from the genome, and consequent loss, of one of the two copies (*Figure 7.1*). Depending upon the precise position at which recombination occurs, either the mutant or the wild-type gene is retained in the genome. In those cases where the mutant gene is retained, a strain is created with the organization of the wild-type chromosome but carrying the mutation brought in with the vector. This phenomenon of allele exchange thus allows the introduction of specific mutations into precisely targeted regions of the yeast genome.

Another, and now more generally used, way of mutating a yeast gene within the yeast genome is shown in *Figure 7.2*. In the simple case shown the aim is to delete a region from within the target gene. Two target gene primers (A and B in *Figure 7.2*) are synthesized that are oppositely orientated and flanking the region to be deleted. Both are 'hybrid' molecules, that is they contain sequences, in their 3′ halves, which also allow them to be used in a PCR reaction with an *E. coli* shuttle vector containing a yeast-selectable marker (*Figure 7.2*). Primers A and B are first

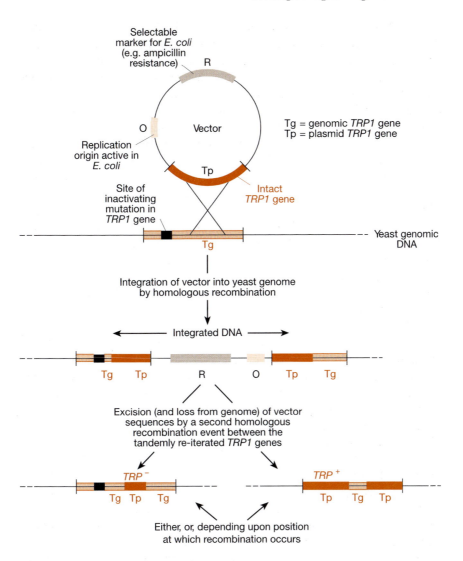

Figure 7.1. Allele exchange using a yeast integrating vector. The vector carrying an intact *TRP1* gene integrates by homologous recombination, producing a tandem duplication of the gene. This is an unstable situation (see *Figure 5.13a*) and so one copy of the gene is excised, again by homologous recombination. Depending upon the position at which crossing-over occurs, a cell will inherit the wild-type gene or the mutant gene. Thus, in a fraction of progeny cells, the original defect will be repaired.

used in a PCR reaction to amplify the yeast-selectable marker and this generates a product in which the yeast sequences A and B flank the marker. When such a PCR product is transformed into yeast cells it usually integrates by homologous recombination, via sequences A and B,

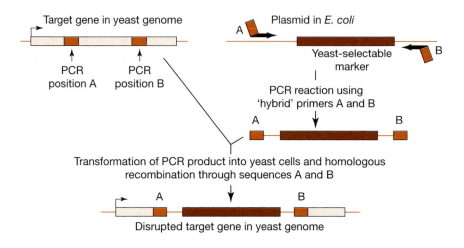

Figure 7.2. Gene manipulation in yeast by homologous recombination of PCR products.

and the target gene fragment is deleted and thereby disrupted at the desired location. This is but one application of this highly versatile method but, by the appropriate selection of primer sequences (i.e. by varying the position within the yeast genome chosen for the design of primers A and B) and/or incorporating additional sequences within the shuttle vector region to be amplified, the yeast genome can be modified almost at will.

7.1.2 Yeast extrachromosomal vectors and YACs

Yeast is an important organism for the commercial production of medically important proteins, such as viral vaccines. In general, the yield of protein is proportional to the number of copies of the gene present within the cell. Integrating vectors are not generally used for the production of heterologous proteins because they are present in only a single copy in the genome, and are rapidly lost by allelic exchange. Extrachromosomal vectors, based upon a naturally occurring plasmid of yeast called the 2μ circle, are commonly used for this purpose [2]. Yeast cells transformed with recombinants generated in 2μ circle-based plasmids contain 50–100 copies of the vector, which is stably maintained within the cell, even in the absence of continued selection.

The class of extrachromosomal vectors of most direct relevance to research in higher eukaryotes are those which allow the cloning of extremely large segments of DNA. Large segments are needed for a variety of purposes, for example, to isolate in one piece the very large genes found in higher eukaryotes or to map a whole genome by establishing ordered arrays of cloned fragments. This can now be achieved in bacteria, for example using BACs (Section 5.3.6), but yeast vectors which permit the formation of yeast artificial chromosomes (YACs) have also been of great value [3]. Eukaryotic chromosomes contain a centromere, which is the site of attachment to the spindle, the structure that pulls the chromosomes apart during mitosis and meiosis. Chromosomes are bounded at each end by a telomere, a region that is essential for complete chromosomal replication. Yeast telomeres and centromeres have been isolated by gene cloning. DNA replication in eukaryotic chromosomes is initiated at multiple sites, scattered along the chromosome. Origins of DNA replication in yeast have been cloned by taking advantage of their ability to confer extrachromosomal replication on circular DNA molecules. They are termed autonomously replicating sequences (ARSs). A typical YAC vector contains a centromere, an ARS, two telomeres and two yeast-selectable markers (*Figure 7.3*). These elements are present in a configuration such that when large fragments of DNA (of up to 500 kb in size) are ligated into them in the manner shown in *Figure 7.3*, a pseudo-chromosome is generated which is stably maintained in yeast as a linear molecule. Such YAC clones can be propagated in yeast in much the same way as a plasmid in *E. coli*. However, for detailed analysis of individual genes it is normally necessary to sub-clone into *E. coli* a fragment from the YAC clone containing the gene of interest.

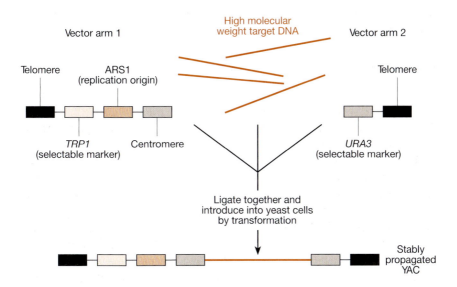

Figure 7.3. The construction of YACs. The target DNA is partially cleaved with a restriction enzyme and ligated to the two vector arms to produce a YAC clone library. It is essential that the starting DNA be of a very large size and the target is sometimes purified by PFGE before insertion, to remove smaller fragments that would otherwise predominate in the library.

7.1.3 The yeast two-hybrid system

Protein–protein interactions are fundamental to the operation of all biological systems and identifying the proteins that a given protein interacts with is immensely important in understanding its function. The yeast two-hybrid system is a widely used method of quantitating interactions between two proteins that are known to interact (*Figure 7.4*). More importantly, it can also be used to clone the cDNAs encoding novel proteins that interact with a chosen target protein.

The yeast two-hybrid system is a conceptually very elegant technique that relies on the fact that transcription factors are modular. They contain a DNA-binding domain and a separate transcriptional activation domain; neither part alone will direct transcription. Consider first the case of two proteins, A and B, that are known to interact with each other. In a typical experiment the cDNA encoding A is cloned in such a way as to yield, in yeast cells, a fusion protein with the DNA-binding domain of the *GAL4* gene (*Figure 7.3*). The cDNA encoding B is cloned in such a way as to yield, in yeast cells, a fusion protein with the *GAL4* trancriptional activation domain. Both fusion proteins are co-transformed into a yeast strain that contains a reporter gene. The promoter of the reporter gene contains a binding site for GAL4 and the reporter gene encodes β-gal. Depending upon their degree of avidity for one another, the A and B fusion proteins will combine together to form a functional transcriptional factor. This factor will direct transcription of the reporter gene and the β-gal protein that accumulates can be detected, either enzymatically or by staining of living cells, just as in bacteria (Section 5.2.1). The amount of enzyme that accumulates can give a measure of the mutual binding affinity of A and B.

The above example illustrates the basic principle. However, the most common use of the two-hybrid system is to take a known gene, again let

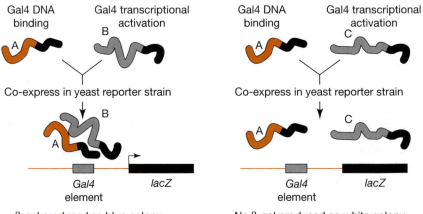

Figure 7.4. The yeast two-hybrid system.

us call it A, and to identify X, Y, Z and any other proteins that A interacts with. This is very much a fishing operation and, appropriately, A is termed the bait and X, Y and Z are termed the prey. The bait is made in the way described above for protein A but the prey constructs derive from a total cell cDNA library. The cDNAs are inserted into the vector, just as described above for protein B, but now each yeast transformant will contain a different fusion protein. There are two different strategies to identify the cDNAs that encode proteins which interact with the bait. The first is that described above, to transform the bait and prey into a strain containing the *GAL4:β-gal* fusion gene. The colony, or colonies, that stain blue on plates containing a hydrolyzable β-gal substrate are picked by visual inspection. Purifying the positive clones will normally entail several rounds of replating and reselection and, to avoid this, an alternative method is often used. The bait and the prey are transformed into a strain that is auxotrophic for leucine because of a mutation in the *leu2* gene. The strain contains within its genome the *leu2* gene under the transcriptional control of GAL4. After transformation with the bait and prey, the strain is plated on nutrient medium lacking leucine. Under this selection the only colonies that grow will be those that express the *GAL4:leu2* fusion gene and this will only be transcribed if the bait and prey molecules bind to each other to yield an active, bipartite transcription factor.

Having identified putative prey molecules a set of genetic screens is used to confirm that the interaction is real and not the result of artifact. Even then it is necessary to be cautious and to confirm that the predicted interaction does indeed occur in the real world, that is in the cell type that the cDNA library was made from. This could, for example, be done by purifying the bait using an antibody and determining whether the predicted prey molecule co-purifies with the bait.

What then are the typical uses of the two-hybrid system? Most commonly, it is used when one has cloned a gene that is involved in a known cellular function and one wishes to identify other proteins that are involved in the process. It can also be used to investigate function. If, for example, a gene is cloned that has no homolog in the databases, or if the homolog that is present has no known function, then identifying the proteins that interact with it could give clues as to its likely role in the cell.

7.2 Transformation of mammalian cells in tissue culture

Yeast is of great value as an organism for basic research and for producing heterologous proteins. For some purposes, however, it is essential to be able to transform mammalian cells. While yeast and mammals share many basic properties, there are important differences in the precise mechanics of gene expression. Mammalian transcription signals are not generally recognized in yeast, mammalian RNA transcripts are not spliced and mammalian proteins are not correctly glycosylated.

Therefore basic research into the control of higher eukaryotic gene expression and the commercial production of some proteins, where secondary modification is considered important, have required the development of transformation methods for mammalian cells. These techniques enable us to investigate the properties and differentiation of mammalian cells representative of most cell lineages in much the same way that single-celled organisms, such as yeast, are studied.

Because they are normally protected from the rigors of the world, mammalian cells are much less robust than yeast cells. They require a carefully balanced growth medium and strict control of pH and temperature. Even when provided with these conditions, cells freshly isolated from a tissue have only a finite lifetime in culture; they will grow and divide for only a limited number of generations. However, in cultures derived from some tissues, mutant cells arise which have acquired effective immortality in culture. These cell lines are used as the host cells for transformation.

The changes which accompany immortalization mimic events occurring when a cell is transformed by an oncogene and, in those cases where cell lines do not arise spontaneously upon prolonged passage, tumors are an important source of novel cell lines. Although tumor cells often lose the ability to enter their normal pathway of differentiation, it is sometimes possible to identify cell lines which will replicate under one set of culture conditions and differentiate under another. For example, cell lines derived from mouse teratocarcinomas, such as the F9 embryonal carcinoma cell, will differentiate to form visceral endoderm (one of the tissues produced early during embryogenesis) when they are deprived of a surface for attachment and exposed to retinoic acid [4].

Where a naturally occurring tumor cell line is not available, it is sometimes possible to generate cell lines by exposure of appropriate primary cultures to oncogenic viruses. Thus, in many genetic studies there is a particular need for a renewable supply of DNA from key individuals. This is especially important in studies of those genetic diseases in which the affected people do not survive for a long period after diagnosis. To overcome this problem, and to avoid resampling, it is essential to establish an immortalized cell line. The Epstein–Barr virus (EBV) is able to integrate into human B lymphocytes and transform them into a permanent cell line. To establish a lymphoblastoid cell line, a mixed population of T and B lymphocytes is exposed to EBV and cultured for 3–4 weeks. EBV transformation does not alter the karyotype, and the cells grow very rapidly and can be stored frozen for extended periods of time.

7.2.1 Transformation of mammalian cells with integrating vectors

Mammalian tissue culture cells readily take up DNA when subjected to electric shock (electroporation), or when the DNA is co-precipitated with calcium phosphate and overlaid on to a monolayer of cells growing on a plastic Petri dish. Depending upon the particular method used to introduce the DNA and the nature of the cell line, a large fraction of cells will take up the vector and transiently express genes contained in it. This phenomenon of transient expression is a useful method of studying gene regulation, because cells can be analyzed very soon after exposure to the DNA. Only a very few cells go on to produce stable transformants, where the introduced DNA has integrated into the chromosome. It is therefore necessary to have some means of selecting for transformant clones. The most commonly used method of selecting a stable transformant is to include within the DNA molecule a gene which confers resistance to a toxic drug, such as the protein synthesis inhibitor G418. This drug is similar in structure to the antibiotic neomycin. A naturally occurring bacterial resistance gene (some-times termed the neo gene) acts by phosphorylating, and hence inactivating, neomycin and G418. In order to direct expression of the neo gene in a mammalian cell, its coding region has been fused to transcription and RNA processing signals which function in eukaryotic cells. A typical G418 resistance vector, pSV2-neo [5], contains a promoter and RNA processing signals from the SV40 tumor virus flanking the drug resistance gene (*Figure 7.5*). If two different DNA molecules are introduced together into cells, then a high proportion of stably transformed clones will contain both sequences. This technique of co-transformation is very useful because, by mixing a DNA sequence of interest with a G418 resistance vector prior to precipitation with calcium phosphate, the DNA can be introduced into a cell without the necessity of first cloning it into a vector.

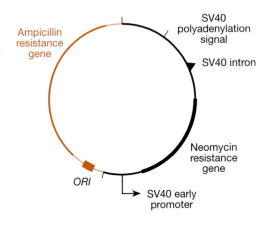

Figure 7.5. A mammalian integrating vector. This prototypic vector contains transcription and RNA processing signals from the animal virus SV40 driving expression of a bacterial neomycin resistance gene. This confers resistance to G418, the related drug to neomycin, that is active in animal cells. It is a shuttle vector for *E. coli* so it also contains an ampicillin resistance gene and an *E. coli* origin of replication.

When mammalian cells are stably transformed with a vector such as pSV2-neo, the DNA integrates at one or more regions within the genome by a process of non-homologous recombination. Homologous recombination does also occur (see Section 7.3.2), but very much less frequently than in yeast, where almost every recombinational event is homologous. The incoming vector DNA may be interrupted at any point along its length, and the sites of integration within the genome also appear to be more or less random. This can pose problems. The metazoan genome is subject to a hierarchy of control not apparent in unicellular organisms. Whole regions of the genome, containing genes which are not expressed in a particular cell type, may be held in an inactive state that is heritably maintained within that cell lineage (see Section 1.8). This repression mechanism can operate to inactivate foreign genes integrated within such regions. Such position effects can be circumvented by using extrachromosomal vectors based upon viral replicons, such as EBV.

7.2.2 Transformation of mammalian cells with retroviral vectors

Retroviral particles contain a genome comprising single-stranded RNA. In infected cells the RNA is copied into a circular, dsDNA which integrates into the cellular genome to form a provirus. The provirus contains long terminal repeats (LTRs), virally derived sequences a few hundred nucleotides long (*Figure 7.6*). Retroviral vectors contain the two LTRs flanking a selectable marker, such as a G418 resistance gene, and other viral sequences necessary for replication and packaging of the RNA into infectious particles. They also contain bacterial plasmid sequences, and DNA to be introduced into a mammalian cell is first cloned into a site between the LTRs using *E. coli* as a host. The recombinant vector is then introduced, by transformation and selection for G418 resistance, into mammalian cell lines that contain the viral genes encoding the proteins necessary to package phage RNA into infectious particles. Because the DNA insert in the vector is flanked by the viral LTRs and packaging sequences, it is transcribed, packaged and exported from the cell. These pseudo-viral particles are then used in a second transformation step and,

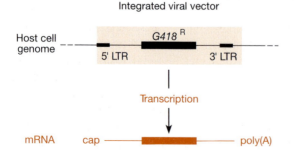

Figure 7.6. A retroviral vector. In such a vector, expression of G418 resistance and integration into the genome are both directed by retroviral DNA sequences.

because viral infection is a highly efficient process, most or all of the cells become transformed. This single step, high efficiency transformation of cells that may be refractory to other methods of introducing DNA, constitutes one of the great advantages of retroviral vectors.

A further advantage of retroviral vectors is that their integration is a relatively controlled process. In contrast to transformation with randomly integrating vectors, which normally yields cell clones containing multiple gene copies, retroviral vectors integrate into the host cell genome as a single copy. Also, the virus always integrates with the LTR sequences at its boundaries (*Figure 7.6*). They are not good vectors for studying tissue-specific expression, because the transcriptional control elements in the LTRs can lead to inappropriate expression of sequences cloned between them. However, they are invaluable if the aim is to obtain constitutive expression.

7.2.3 Regulated expression of cloned genes in mammalian cells

Powerful viral promoters, such as the CMV promoter, are often used to express a cloned gene in mammalian cells when a high constitutive level of expression is to be achieved. However, in order to study the effect of a specific protein on the cells' behavior, regulatable expression under the control of an inducible or repressible promoter is generally used: the protein can then be expressed at will and the consequences for the cell determined. The tet OFF (*Figure 7.7*) and tet ON systems are two commonly used regulatable expression systems and, as their names imply, they respectively allow for repressible and inducible gene expression. One further, very valuable, feature of these methods is that regulation of gene expression is exerted by addition of the anti-bacterial drug tetracycline. Tetracycline has no discernible effect on mammalian cells, so there are no complications due to non-specific effects of the regulatory molecule (as might be the case if a mammalian regulatory system was utilized).

In bacterial cells, the tetracycline repressor protein, tetR, binds to the operator sequences of the tetracycline operon, preventing transcription. In the presence of tetracycline tetR dissociates from the DNA, allowing transcription to occur. In both the tet OFF and tet ON systems, the gene to be expressed (the target gene) is cloned downstream of a hybrid promoter that contains a binding site for tetR and the TATA box and cap site of CMV (*Figure 7.7*). The mammalian cells contain, elsewhere within their genome, the genes encoding one of two alternative regulatory proteins. In the tet OFF system the regulatory factor (tetR/VP16) is a fusion protein carrying the DNA recognition and binding domain of tetR joined to the transcriptional activation domain of the mammalian transcription factor VP16. In the absence of tetracycline tetR/VP16 binds upstream of the target gene and directs its expression. When tetracycline is added to the cells it binds to the tetR/VP16 protein. This causes tetR/VP16 to dissociate from the promoter and transcription ceases. There is a

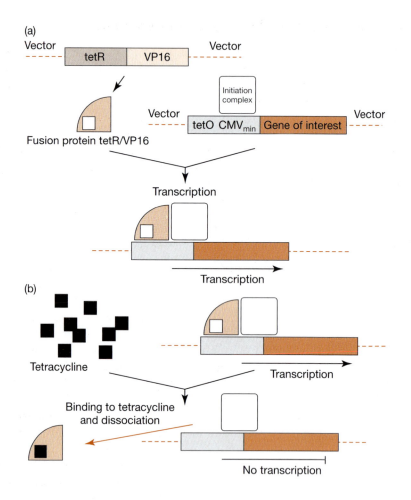

Figure 7.7. The tet OFF system for regulated gene expression. (a) Two constructs are introduced into the cell. One (left) expresses a regulator protein (tetR/VP16) containing tetR protein sequences fused to the VP16 transactivation domain. The second (right) contains the gene of interest (target gene) under the transcriptional control of the tetR/VP16 fusion protein. In the absence of tetracycline the minimal promoter elements within the CMV-derived part of the promoter bind basal transcription factors and the tetR/VP16 fusion protein binds to the tet operator sequences. Together they form an active initiation complex. (b) When tetracycline binds to tetR/VP16, it dissociates from the promoter and transcription ceases.

mutant version of tetR (tetR^m) that acts in exactly the opposite way, that is it binds to the regulatory element only in the presence of tetracycline. Thus by transforming the target gene construct into cells expressing tetR^m/VP16 the target gene can be placed under positive regulatory control by tetracycline. This is known as the tet ON system.

7.3 The creation and use of transgenic mice

Cultured mammalian cells are ideal for many purposes in cell and developmental biology, and great advances have been made using them. Sometimes, however, it is essential to study the whole organism. This is obviously necessary for those genetic diseases where a lesion affects many tissues and organs. It also holds true when the behavior of cells in culture does not faithfully reflect the behavior of the analogous cells *in vivo*. Mammalian development is a complex process, involving multiple interactions of cell with cell and of cell with substratum. It is perhaps not surprising, therefore, that development *in vitro* sometimes fails completely to mirror development *in utero*.

7.3.1 Generation and analysis of transgenic mice

Newly fertilized eggs are collected, prior to fusion of the male and female pronuclei. A tiny volume of DNA solution is sucked up into a very fine glass pipet and injected into the male pronucleus, which can be distinguished from the female pronucleus because it is smaller in size. The eggs are then implanted into pseudo-pregnant foster mothers, generated by mating with vasectomized males. The procedure is technically demanding but, in skilled hands, a proportion of injected eggs survive to term and will contain the injected DNA. Progeny mice are tested for the presence of this DNA by removing a portion of the tail and performing PCR analysis. Integration of the DNA into the genome of cells within the embryo is a poorly understood process, but it seems to occur early during embryonic development. A proportion of the animals resulting from injection are mosaic; the various tissues within the animal contain differing amounts of the injected DNA, and it is absent from some tissues. It is normally necessary to establish a line of mice containing the gene of interest in all tissues. Fortunately, injected DNA integrates into the germ cells in a high proportion of mice. By mating such mosaic mice to non-injected animals and analyzing tail DNA of the progeny, it is possible to identify those containing the integrated gene. The parent in such transgenic lines of animals will pass on the injected gene in a simple Mendelian fashion (*Figure 7.8a*), and is said to display germ-line transmission. The DNA is injected as a linear fragment which forms head to tail concatemers that normally integrate at apparently random locations within the genome. Again, the gene may be susceptible to position effects, so that different mouse lines containing the same gene may show very different levels of gene expression.

Transgenic mice have many important uses. They can be employed as an assay system, to delineate the DNA sequences necessary for tissue-specific gene expression in those cases where there is not an appropriate tissue culture. They can be used to generate mouse models of genetic diseases. If, for example, an oncogene is placed next to a promoter which directs tissue-specific gene expression, it can lead to the formation of

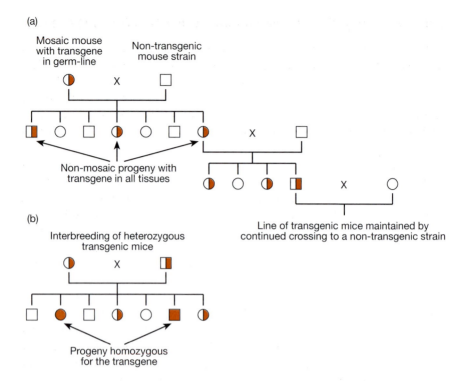

(a)

Mosaic mouse with transgene in germ-line

Non-transgenic mouse strain

Non-mosaic progeny with transgene in all tissues

Line of transgenic mice maintained by continued crossing to a non-transgenic strain

(b)

Interbreeding of heterozygous transgenic mice

Progeny homozygous for the transgene

Figure 7.8. Maintenance and analysis of transgenic mouse strains. (a) This figure shows how a line of transgenic mice can be maintained by crossing to a non-transgenic strain and monitoring progeny for the presence of the transgene to determine which mice are carriers. (b) If the transgene is recessive, its effect can be determined by mating together two transgenic mice. Assuming that the transgene is not essential for fetal development, 25% of the progeny will be homozygous for the transgene.

specific tumors. They can also be used to perform random insertional mutagenesis to search for genes important in mouse development. In a surprisingly high proportion of cases integration of the transgene occurs in a region of the genome which inactivates an important gene. Such a mutation (if recessive) will not manifest itself in a transgenic line maintained by mating to a wild-type mouse (*Figure 7.8a*), because the normal allele will always be present and will provide the missing function. It can, however, be revealed by crossing two sibling mice that are heterozygous for the transgene (*Figure 7.8b*). One-quarter of the progeny of such a cross will be homozygous for the mutation and if, as is often the case, it is a gene required during development, these embryos will die *in utero*. The beauty of this technique of mutagenesis is that the vector DNA used to generate the transgenic mouse will normally be located somewhere in the vicinity of the mutated gene. It is therefore a relatively straight-forward matter to prepare a genomic clone bank from mutant mice, to screen for the integrated DNA, and hence to clone the essential gene.

7.3.2 Gene targeting by homologous recombination in the mouse

Cloning has allowed the isolation of mammalian genes involved in a wide spectrum of biological processes. Even though the biochemical activity or structural role of the gene product will normally be known prior to cloning, it is often very valuable to determine the phenotypic consequences that result if it is absent or mutated. In lower eukaryotes, such as yeast, this has been routine for many years and such techniques are now also possible in mammals. The technique is much more time-consuming and demanding than in yeast because the frequency of homologous recombination in mammalian cells is orders of magnitude lower than in yeast, and because mammals are diploid organisms. The latter problem could, in principle, be overcome by performing disruption in transgenic mice, mating together two transgenic animals and identifying the 25% homozygous mutant progeny (*Figure 7.8b*). However, the frequency of homologous recombination in transgenic mice is probably as low as that observed for mammalian tissue culture cells, in the order of one homologous event to 1000 non-homologous events. Thus it would be totally impractical to generate sufficient numbers of transgenic mice to identify homologous recombinants because of the length of time required. The solution to this problem is to produce mutant mice by creating chimeras with embryonal stem (ES) cells, rather than by direct micro-injection of DNA into eggs [6,7].

ES cells are embryo-derived cells that are pluripotent, that is, they have the ability to participate in normal development and to differentiate into every mouse cell type. If ES cells are introduced into an embryo, a chimeric animal is produced. This is a mosaic, containing both ES and normal cells. Since the germ cells will also be a mixture of ES and normal cells, chimeric animals produce a fraction of offspring derived from ES gametes. ES cells are susceptible to DNA-mediated transformation in culture, using integrating vectors, and such transformed cell lines retain the ability to chimerize with normal cells. This provides an alternative method for generating lines of transgenic mice (*Figure 7.9*).

The importance of the procedure is that hundreds or thousands of ES cell lines, each the product of a different transformation event, can readily be generated *in vitro*. These can then be screened, by PCR, to identify the rare cases in which the gene has integrated by homologous recombination (*Figure 7.10*). There are also now genetic methods to increase the frequency of cell clones in which homologous recombination has occurred [8]. Such a cell line will not display a mutant phenotype which is recessive because the cells are diploid and so retain a normal copy of the gene on the non-disrupted chromosome. However, the effect of homozygosity for the mutation can be determined by breeding homozygotes from the original heterozygous transgenic mice.

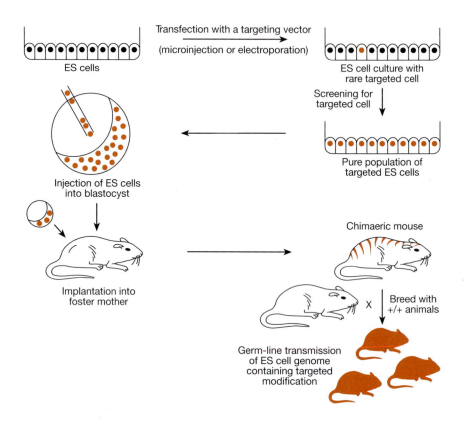

Transfection with a targeting vector
(microinjection or electroporation)

ES cells

ES cell culture with
rare targeted cell

Screening for
targeted cell

Pure population of
targeted ES cells

Injection of ES cells
into blastocyst

Chimaeric mouse

Implantation into
foster mother

Breed with
+/+ animals

Germ-line transmission
of ES cell genome
containing targeted
modification

Figure 7.9. Generation of transgenic mice using ES cells. This is an alternative route to producing a transgenic mouse. Since ES cells are cultured cells which can undergo spontaneous mutation during passage, it is essential that the cells be carefully maintained, otherwise they will not contribute to the germ line of the progeny mice.

7.3.3 Transgenic mouse models of human disease

A good example of the use of targeted gene disruption is provided by a study of the phenotypic effect produced by inactivating the mammalian *Hox-1.5* gene [8]. This same study also illustrates how mice can be used to create a model of a human disease.

Homeotic genes were first identified in the fruit fly *Drosophila melanogaster* as genes which, when mutated, cause aberrant development. The best characterized of these genes, those which help to specify the pattern of differentiation along the antero-posterior axis, are transcription factors. They determine the identity of the various segments by activating genes that direct formation of the structures characteristic of each segment. Mutation in various of these genes can yield remarkably aberrant adult flies with, for example, an additional winged segment,

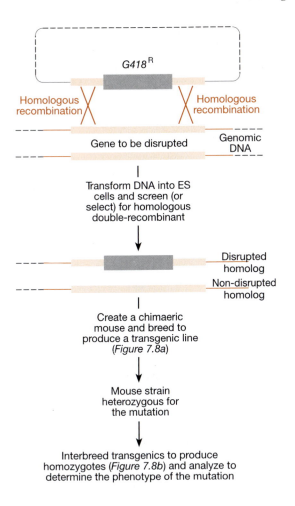

Figure 7.10. Homologous gene disruption in ES cells. A series of cloning steps in *E. coli* are used to generate a construct in which the 5′ proximal and 3′ proximal parts of a gene flank a *G418* resistance gene. If this inserts by homologous recombination, one of the allelic copies of the gene is inactivated. The phenotype of the mutation is then determined by creating a transgenic mouse and breeding to create a homozygote.

giving four wings instead of the normal two, or with a leg-bearing segment on the head. The main *Drosophila* homeotic genes are organized into two clusters and for reasons which are not yet clear, those genes which are active in the most anterior segments of the fly are located nearer the 3′ end of the clusters, and genes that are active in the more posterior segments are located nearer the 5′ end. There is an almost perfect inverse correlation between position of a gene relative to the 5′ end of its respective cluster and its domain of expression within the fly (*Figure 7.11*).

The DNA-binding domain of homeotic genes, the homeobox, is present, in a more or less conserved form, in all metazoan organisms so far studied. Homeobox genes (homeogenes) have been isolated from several different mammals, including man, and, just as in the fruit fly, they play a central role in regulating formation of the antero-posterior axis during embryonic development. Remarkably, given the enormous evolutionary separation and radically different body plans of mammals and fruit flies, homeogenes that specify particular segments in flies have

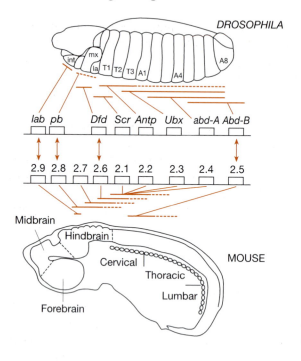

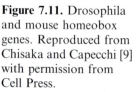

Figure 7.11. Drosophila and mouse homeobox genes. Reproduced from Chisaka and Capecchi [9] with permission from Cell Press.

counterparts in mammals. In the mouse there are four homeogene clusters, found on separate chromosomes, that contain homologous genes in equivalent relative positions. These are believed to have arisen by multiple duplications during evolution from a single cluster, with a structure similar to that obtained by aligning the two *Drosophila* clusters end to end (*Figure 7.11*). Part of the evidence for this hypothesis is that almost every mammalian homeogene is in an identical relative position to its nearest *Drosophila* homolog.

The sequence homology, and conservation of relative order along the chromosome, between *Drosophila* and mammalian homeogenes most probably reflects conservation of function. While adult mammals are not externally segmented, during embryonic development the hindbrain is visibly and functionally divided into segments called rhombomeres. The domains of expression of the mammalian homeobox genes in the rhombomeres of the embryonic brain are related to their positions within the gene cluster, in a very similar fashion to their *Drosophila* counterparts. This observation adds confidence to the idea that the homeogene clusters have been conserved during evolution and are playing fundamentally similar roles in flies, mice and men. The downstream target genes of these and other homeobox genes, which are believed to direct processes such as limb formation, are sufficiently different to ensure that men have arms and flies have wings.

Drosophila can be bred in enormous numbers so that very sophisticated genetic techniques can be used to analyze its development. These are not easily applicable to the mouse and, whilst conservation of gene order, sequence and pattern of expression are strongly indicative of functional homology, they do not constitute definitive proof that the mammalian homeobox genes play analogous roles to those of *Drosophila*. This proof has come from the targeted disruption of the mouse homoebox gene *Hox-1.5* [8]. This was achieved by transfecting into ES cells a construct of the form shown in *Figure 7.10*. It contained a G418 resistance gene, flanked on each side by sequences derived from the *Hox-1.5* gene. This generated a number of G418-resistant cell lines, each originating from a different transformation event. Individual clones were then screened to find one in which two homologous recombination events had occurred, as shown in *Figure 7.10*. The effect of this double crossover was to replace the *Hox-1.5* gene on one of the two homologous chromosomes with an interrupted, inactive copy of the gene. These cells were introduced into a mouse embryo to create a chimera which was then re-implanted in a pseudo-pregnant mother.

The ES cells which were used to create the disruptant contained a mutation in the agouti coat color gene, so that chimeric mice were readily identified. Breeding from these chimeric animals yielded mice which were entirely agouti in coat color, showing that some of the re-implanted ES cells had contributed to the germ line. By inter-crossing these animals and analyzing the genomic DNA of the progeny by Southern transfers, mice homozygous for the disrupted *Hox-1.5* gene were obtained. These embryos displayed severe defects including athymia and aparathyroidy, and died a few hours after birth. A more detailed investigation revealed that the pattern of malformation closely resembled that observed for human Di-George syndrome. Di-George syndrome itself is probably not caused by disruption of the human equivalent of *Hox-1.5*, because it maps to the wrong chromosomal location, but clearly the pathological sequence has many features in common with events in the transgenic mice.

The phenotype obtained by disrupting the *Hox-1.5* gene is complex, but it can be partially correlated with the pattern of *Hox-1.5* expression. Redundancy of function between the homologous *Hox* genes may explain why the correlation is not simple. When *Hox-1.5* is disrupted, another homeobox gene may be able to fulfill some of its functions. It probably also reflects the fact that disruption is a very unsubtle form of genetic alteration. A more informative mutation would be one that altered the pattern of expression of a homeobox gene, such that it was expressed in regions of the embryo where it was normally silent. Such ectopic expression can be obtained by linking the coding region of a gene to a promoter from another gene to create a gene fusion.

When the mouse *Hox-1.1* gene is coupled to a constitutively expressed actin promoter it produces a remarkable phenotype [10]. The mice have an extra cervical vertebra and variations in the most anterior units, consistent with a posterior to anterior transformation. This change in vertebrate body plan, produced by ectopic expression of the mouse homolog of a *Drosophila* homeotic gene, provides a graphic illustration of the fundamental similarity in the body-building processes between different metazoan organisms.

Another excellent example of the interplay between fly, mouse and human genetics is provided by the *Pax* genes. This class of genes was originally identified in the *Drosophila* mutant paired. Using the DNA-binding domain (paired box) of the paired gene as an initial probe, a family of *Pax* genes has been isolated from the mouse. One of the genes, *Pax-3*, was found to be expressed in the developing nervous system of the mouse embryo and located in the proximal portion of mouse chromosome 1. A mouse mutation, *Splotch*, maps to this approximate region and has defective function of the embryonic neural crest. It has now been shown that *Pax-3* is the *Splotch* gene [11]. A mutation in another *Pax* gene, *Pax-6*, causes the small-eye mutation in mice [12]. Waardenburg's syndrome (WS1) is an autosomal dominant condition in man characterized by deafness and changes in pigmentation. These symptoms point to defective function of the embryonic neural crest. WS1 has been mapped to a region of chromosome 2 that is homologous to the region of mouse chromosome 1 containing the *Splotch* gene. The human homolog of the *Pax-3* gene has now been cloned, and sequence analysis shows that there are mutations in this gene in patients affected with WS1 [13,14]. Thus, *Splotch* and the gene for WS1 are homologous *Pax* genes.

7.4 The impact of whole genome DNA sequence determination: proteomics and micro-arrays

The discovery of the homeobox and the paired-box genes illustrates the enormous power of having a genetic map and human geneticists now know the entire sequence of the human genome. It is important to remember, however, that a complete genome sequence is not a functional genetic map. A major fraction of the human genome will be composed of computer-predicted genes that have no known function (Section 3.3). Studies from model systems, such as yeast and *Drosophila*, will be essential in ascribing a function to such genes. However, the various genome sequencing projects will have an enormous impact on the way that genes are isolated and manipulated. One immediate effect will be to reduce the need for many of the standard cloning and gene analysis techniques described in this book. When the DNA sequences of all the genes are known, then PCR can be used to isolate and manipulate the desired DNA sequences using genomic DNA as a template. It will still, however, often be necessary to initially identify the gene of choice and here a knowledge of the genome sequence will also prove a major benefit.

One extremely important consequence of a whole genome sequence will be to make the screening methods described in Chapter 6 in some cases redundant: if a short region of protein sequence can be determined for the gene of interest, then the database of genome sequence can be searched for a match. Here developments in DNA sequencing speed, that have allowed the entire genome sequence to be determined, have been paralleled by improvements in the speed and sensitivity of protein analysis using mass spectrometry. Tiny samples of protein, of the amount that can be obtained from one spot on a two-dimensional protein gel, can be used for the partial sequence analysis. This combination of techniques forms the basis for a new approach to studying differential gene expression that is sometimes termed functional proteomics. Two closely related tissue samples, for example from a normal and from a diseased cell, are analyzed on a two-dimensional gel. The protein spots that are common to the two samples provide a framework for alignment of the gels. Those that show a difference are partially sequenced by mass spectrometry and the genes that encode them are identified from the database.

The genome project also forms the basis for an RNA analysis method that holds enormous promise for disease diagnosis. DNA micro-array analysis is conceptually similar to the differential screening of cDNA libraries (Section 6.2.1) but the cDNAs are immobilized as many thousands of tiny spots on a glass slide. Each spotted cDNA is, most often, a PCR-generated copy of the entire insert (see Section 6.1.3) present within the MCS of a cDNA clone. However, if a cDNA is not available, such as would be the case for a computer-predicted gene, genomic DNA can be used as a template for PCR with suitable primers.

Using one or other approach (i.e. a cDNA or a genomic template) micro-arrays are prepared bearing DNA from some or all of the genes in the genome. The two mRNA populations to be compared are separately used as templates for cDNA synthesis, such that one of the resultant cDNA probes is fluorescently labeled with one dye (labeled 'red' in *Figure 7.12*) and the other is fluorescently labeled with a different dye (labeled 'green' in *Figure 7.12*). The two labeled cDNA probes are then mixed together and hybridized to the micro-array slides. The number of labeled probe molecules that hybridize with an individual spot in the micro-array is a measure of the relative abundance of that particular probe species (Section 6.2.1). Since cDNAs from two different mRNA populations are present in the hybridization, each spot has the potential to hybridize to each mRNA and this will result in a particular ratio of red to green dye being localized at this spot. A spot where there is hybridization only to the red probe will of course appear red when the micro-array slide is placed on an optical scanner. Similarly, a spot where there is hybridization only to the green probe will appear green but the co-incidence of a red and green signal will, because of the optical physics involved, cause a spot where there is hybridization to both probes to appear yellow. This color and spot intensity information can be used to build up an expression profile showing changes in gene expression between different tissues.

As with functional proteomics, it is possible to use this method to compare any two biological or clinical samples. For example, tumor

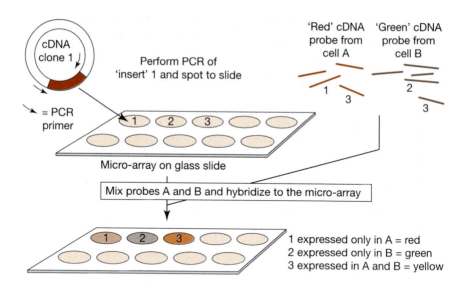

Figure 7.12. DNA micro-array analysis. In this diagrammatic representation just three spots, showing the three possible patterns of hybridization, are represented. In a full-scale experiment many thousands of different cDNAs might need to be amplified and spotted onto the slide. Also, the DNA for micro-arraying need not, as shown here, always derive from a cDNA clone (see text).

samples and a control sample can be used to determine which genes are aberrantly expressed in particular forms of cancer. This will be of great fundamental importance, since it will indicate which gene products are promising targets for devising a therapy. It is also of immediate clinical importance, because it will allow for tumor typing in those situations where histological discrimination may be difficult or impossible. The uses extend far beyond cancer into all aspects of human disease diagnosis.

References

1. **Struhl, K., Cameron, J.R. and Davis, R.W.** (1979) High-frequency transformation of yeast: autonomous replication of hybrid DNA molecules. *Proc. Natl Acad. Sci. USA* **16**: 1035.
2. **Beggs, J.D.** (1978) Transformation of yeast by a replicating hybrid plasmid. *Nature* **175**: 104.
3. **Burke, D.T., Carle, G. and Olson, M.** (1987) Cloning of large segments of exogenous DNA into yeast by means of artificial chromosome vectors. *Science* **236**: 806.
4. **Hogan, B.L.M., Taylor, A. and Adamson, E.** (1981) Cell interactions modulate embryonal carcinoma cell differentiation into parietal or visceral endoderm. *Nature* **291**: 235.
5. **Southern, P.J. and Berg, P.** (1982) Transformation of mammalian cells to antibiotic resistance with a bacterial gene under control of the SV40 early region promoter. *J. Mol. Appl. Genet.* **1**: 327.

6. **Gossen, M., Freundlieb, S., Bender, G., Muller, G., Hillen, W. and Bujard, H.** (1995) Transcriptional activation by tetracyclines in mammalian cells. *Science* **268**: 1766–1769.
7. **Martin, G.R.** (1981) Isolation of a pluripotent cell line from early mouse embryos cultured in medium conditioned by teratocarcinoma stem cells. *Proc. Natl Acad. Sci. USA* **78**: 73–64.
8. **Evans, M.J. and Kaufman, M.H.** (1981) Establishment in culture of pluripotential cells from mouse embryos. *Nature* **292**: 154.
9. **Chisaka, O. and Capecchi, M.** (1991) Regionally restricted developmental defects resulting from targeted disruption of the mouse homeobox gene hox-1.5. *Nature* **350**: 473.
10. **McGinnis, W. and Krumlauf, R.** (1991) Homeobox genes and axial patterning. *Cell* **68**: 283.
11. **Kessel, M., Balling, R. and Gruss, P.** (1990) Variations of cervical vertebrae after expression of a Hox-1.1 transgene in mice. *Cell* **61**: 301.
12. **Epstein, D.J., Vekemans, M. and Gros, P.** (1991) Splotch (Sp2H), a mutation affecting development of the mouse neural tube, shows a deletion within the paired homeodomain of Pax-3. *Cell* **67**: 767.
13. **Hill, R.E., Favor, J., Hogan, B.L.M., Ton, C.C.T., Saunders, G.F., Hanson, I.M., Prosser, J., Jordan, T., Hartie, N.D. and Heyningen, V.** (1991) Mouse small eye results from mutations in a paired-like homeobox-containing gene. *Nature* **354**: 522.
14. **Tassabehji, M., Read, A.R., Newton, V.E., Harris, R., Balling, R., Gruss, P. and Strachan, T.** (1992) Waardenburg's syndrome patients have mutations in the human homologue of the Pax-3 paired box gene. *Nature* **355**: 635.
15. **Baldwin, C.T., Hoth, C.F., Amos, J.A., da-Silva, E.O. and Milunsky, A.** (1992) An exonic mutation in the HuP2 paired domain gene causes Waardenburg's syndrome. *Nature* **355**: 637.

Further reading

Brent, R. and Finley, R.L. Jr (1997) Understanding gene and allele function with two-hybrid methods. *Annu. Rev. Genet.* **31**: 663–704.
Capecchi, M.R. (1989) The new mouse genetics: altering the genome by gene targeting. *Trends Genet.* **5**: 70–76.
Granjeaud, S., Bertucci, F. and Jordan, B.R. (1999) Expression profiling: DNA arrays in many guises. *Bioessays* **21**: 781–790.
Hill, R. and Van Heyningen, V. (1992) Mouse mutations and human disorders are paired. *Trends Genet.* **8**: 119–120.
Kingsman, S.M. and Kingsman, A.J. (1988) *Genetic Engineering.* Blackwell Scientific Publications, Oxford.
Krumlauf, R. (1991) The Hox gene family in transgenic mice. *Curr. Opin. Biotechnol.* **2**: 796–801.
McGinnis, W. and Krumlauf, R. (1991) Homeobox genes and axial patterning. *Cell* **68**: 283–302.
Schlessinger, D. (1990) Yeast artificial chromosomes; tools for mapping and analysis of complex genomes. *Trends Genet.* **6**: 255–258.
Sedivy, J.M. and Joyner, A.L. (1992) *Gene Targeting.* W.H. Freeman, London.

Index

2 + 4 rule, 25
2μ circle, 122

α-peptide, 83
Activation domain, 16
Adenine, 1
Adenosine, 1
Affinity chromatography, of RNA, 104
Agouti coat color, 137
Allellic variation, 42
Amino acid, 6
Aminoacyl-tRNA synthetase, 6
Amplification of DNA sequences, in PCR, 61
Amplification refractory mutation detection system (ARMS), 67–68
Antibiotic resistance, 82
Anticodon, 6
Aparathyroidia, 137
ARMS, 67–69
ARSs, 123
Athymia, 137
AU-rich sequence, 17
AUG, 6
Autonomously replicating sequences (ARSs), 123
Autoradiography, 27
Auxotrophic strains, 120

β-galactosidase, screening method, 83
BACs, 123
Bacterial electroporation, 87

Bacterial transformation, 87
Bait, 125
*Bam*HI, 36
Bases, 1
Blunt termini, 37

Cap site, 57
CCM, 71, 74
cDNA, 56, 103
    double stranded, 104
    expression, in mammalian cells, 114
    methylation, 37
    screening method, 87
    start, 10
    synthesis, 104
Cell immortalization, 126
Chain terminators, 48
Chemical cleavage of mismatches (CCM), 71
Chimera, 137
Chromatin, transcriptionally inactive, 18
Chromosome, condensation, 19
    walking, 94, 115
Clone bank, 90
Cloning, PCR products, 88
CMV promoter, 129
Co-transformation, 127
Codon, 6
Compatible overhanging termini, 85
Computer use in the search of coding regions, 58
Cos sites, 90

Cosmid, 94
Cosmid library, 97
CpG dinucleotide, 36, 59
CpG islands, 59, 115, 116
Crossing over, 34
   unequal, 42
CTF, 15
Cystic fibrosis, 67, 114
Cytidine, 1
Cytosine, 1

Deletion series, use of in DNA
      sequencing, 53
Di-George syndrome, 137
Differential display, 108, 111
Differential screening, 106, 111
Directional cloning, 87
Disease model, 134
DMPK gene, 40
DNA, 1
   base pairing of, 1
   chemical synthesis, 28
   circular, 1
   closed circular, 2
   complementary (cDNA), 103
   content, of a human cell, 18
   content, of an *E. coli* cell, 18
   double helix, 1
   double stranded, 1
   genomic, analysis, 58
   genomic, as target in genomic
         libraries, 97
   genomic, restriction of, 103
   heteroduplex, 72, 74
   homoduplex, 72
   linear, 1
   looping out, in transcriptional
         control, 16
   micro-array analysis, 139, 140
   minipreparation, 88
   mitochondrial, 2
   non-coding, 17
   replication, 1
   sequencing, 47–55
      artifacts, 53, 55

   automated, 50
   by chain termination, 48
   chemical degradation, 47
   Maxam and Gilbert, 47, 74
   shotgun, 55
   single stranded vs. double
         stranded template, 52
   sub-cloning, 85
   supercoiled, 2
   target, in PCR, 60
DNA-binding domain, 16, 124
DNA database, 59
DNA ligase, 1, 87
DNA polymerase, 1, 29
DNA polymerase I, 30
DNA:DNA duplex, 25
DNA:RNA duplex, 25
Domain control region (DCR), 20
Dynamic mutations, 39, 74

E. coli, chromosome, 9
   genome, 2
EBV, 126
*Eco*RI methylase, 106
Electrophoresis, in agarose gels, 25
   in polyacrylamide gels, 25
   pulsed field, 26
Electroporation, 126
Embryonal stem (ES) cells, 133
Embryonic development, 135
Endonucleases, 20
Enhancer binding protein, 15
Enhancers, 15
Epstein-Barr virus (EBV), 126
ES cells, 133
   transfection, 137
   transformation, 133
Ethidium bromide, 26
Eukaryote, 9
Exon, 17
Expression screening, 111
Expression vector, 112
Extracellular matrix, 17

Factor VIII, 40, 41

Ferritin mRNA, 13
Fibronectin, 17
FISH, 34
Fluorescein, 34
Fluorescent labeling, in automated
    DNA sequencing, 51
Forced cloning, 87
Forensic medicine, 44
Formamide, 25
Fragile X syndrome, 74
Friedreich's ataxia, 74
Functional proteomics, 139

G418 resistance, 127, 128
GAL4, 124
Galactoside, 112
Gene cloning, 56, 81
Gene expression, cell-type specific, 15
    in mammalian cells, 129
    post-transcriptional regulation,
        13, 15
Gene library, 90
Gene tracking, 66
Genetic analysis, pre-natal, 66
Genetic code, 6
Genetic disease, diagnosis of, 44
    mouse models of, 131–138
Genetic linkage analysis, 34
Genomic library, 86
Globin genes, 20
Golgi complex, 8
Guanine, 1
Guanosine, 1

Helper phage, 93
Hemophilia A, 40
Histones, 18
Homeobox, 135
Homeobox gene disruption, 137
Homeotic genes, 135
Homologous recombination, 99–
    100, 121
Hox 1.5, 134
Human genome project, 59, 66
Huntington disease, 74

Huntington gene, 74
Hybridization, *in situ*, 33, 34, 97
    in situ, fluorescent (FISH), 34
    in solution, 24
    inter-specific, 115, 116
    kinetics, 24
    of immobilized target, 24
    of nucleic acid, 23
    rate, 24
    stringency, 25
    target, 24
    probe, 23, 34
Hydroxyapatite, 108

Identity testing, 42
Immunodetection, of probe, 30
In vitro packaging, 90, 93, 97
Inducible promoter, 129
Initiation codon, 6, 8
Intron, 17
IRE, 14
IRE-BP, 13, 14
Iron response elements, 13

Klenow fragment, 30
Kozak sequence, 12, 59

λ bacteriophage, 89
    arms, 91
    cos sites, 90
    genome, 90
    insertion vector, 92
    packaging, 90
    replacement vector, 91
λgt10, 92
λgt11, 112
λ phage library, 97
λZAP, 92, 93, 116
lac operon, 9, 83
lac repressor, 10
lacZ, 83
lacZ fusion protein, 112
Library, amplified, 97
    lifting, 97
Ligase, thermostable, 69

Linearization, 85
Linkage analysis, 66
Linkers, 105
Locus control region (LCR), 20
Logarithm of the odds (Lod), 34
Long terminal repeats (LTRs), 128
LTRs, 128

M13 bacteriophage, 85, 89
  helper phage, 85, 93
Maxam and Gilbert method, 47, 74
MCS, 82
Melting temperature (Tm), 23
Membrane fusion, 33
Metallothionein, 16
Micro-array analysis, 139
Microsatellite repeats, 41
Microsatellites, 66
  di-nucleotides, 65
  tetra-nucleotides, 65
  tri-nucleotides, 65
Mosaic, 131
MRNA, 5, 15
  3' end, 11
  3' non-coding, 12
  3' terminus, 17
  5' cap, 11
  5' end, 11
  5' non-coding, 12
  cytoplasmic, 11
  differential display, 108
  eukaryotic, 11
  globin, 103
  hybridization, 139
  mapping, 63
  poly(A) tail, 11, 12, 104
  precursor, 11, 17
  prokaryotic, 11
  quantitation of, 58
  stability, 13
Multi-allelic variation, 41
Multi-cloning site (MCS), 82
Myotonic dystrophy, 40, 74

Neomycin, 127

Neurofibromatosis type 2 (Nf2), 70
Nitrocellulose filters, DNA binding
  to, 28
Northern blot, 23, 28, 33
*Not*I, 37
Nuclear localization signal, 8
Nuclease protection, 56
Nuclease S1, 57
Nucleoside, 1
Nucleosome, 18, 19
Nucleotide, 1
Nylon membranes, DNA binding
  to, 28

Okazaki fragments, 1
OLA, 68–69
Oligo ligation assay (OLA), 68–69
Oligo-dT, 104
Oligonucleotide, hybridization, 111
  labeling, 29
Oligonucleotides, as primers, 60
  as probes, 25
  mixed, 112
Open reading frame (ORF), 56
Operons, 9
ORF identification, 115

P1 bacteriophage, 95
Paired box, 138
Paired gene, 138
Palindromic sequences, 36
Ppaternity testing, 42, 44
Pax genes, 138
pBluescript, 82,87,93
  preparation of single stranded, 85
pBR322, 82
PCR, 60–64
  diagnostic, in AIDS, 78
    in chlamydiae infection, 78
    in cytomegalovirus (CMV)
      infection, 78
    in hepatitis B virus (HBV)
      infection, 77
    in herpes virus infection, 78
    in infectious diseases, 76

error rate, 63
false positives, 61
in human diseases diagnosis,
65–78
Phosphatase treatment, 87
Piperidine, 74
Poly(A) polymerase, 17
Poly(A) tail, 11, 56
in mRNA purification, 104
Polyacrylamide gel, denaturing, in
DNA sequencing, 49
Polyadenylation signal, 59
Polyethylene glycol, promoting
membrane fusion, 33
Polylinker, 82
Polymerase, slippage, 42
Polymerase, stuttering, 42
Polymerase chain reaction (PCR),
60–63
diagnostic, 38
Polysome, 7
Pore size, acrylamide vs. agarose,
26
Position effects, 131
Positional cloning, 114
Pre-natal diagnosis, 116
Prey, 125
Primer extension, 56, 57
Probe, labeling, 29
non-radioactive, 33
Prokaryote, 9
Protein, functional domains, 17
post-translational processing, 8
Protein database, 59
Proteomics, 139
Provirus, 128
Pseudo-pregnant foster mothers,
131, 137
pSV2-neo, 127, 128
pUC19, 82

RACE-PCR, 56, 63
Rapid amplification of cDNA ends
(RACE), 56
Recombination, 120

homologous, 128
homologous, in transgenic mice,
133
non-homologous, 128
Repressible promoter, 129
Restriction enzyme, mapping,
34–39
methylase, 105
recognition sequences, 36
Restriction fragment length
polymorphism (RFLP), 41
Restriction mapping, of genomic
DNA, 38
Reverse transcriptase, 56, 104
Reverse transcription, 109
RFLPs, 41, 65, 115
Rhodamine, 34
Ribose, in RNA, 2
Ribosome, 7
binding site, 10
RNA, 3
antisense, 84
messenger, 3
packaging, 128
processing signals, 127
ribosomal, 3
sense, 84
structural, 3, 6, 15
total cellular, 5, 104
transfer, 3
RNA polymerase, 10, 15
T3 promoter, 83
T7 promoter, 83
I, 15
II, 15, 16
III, 15
RNAase inhibitors, 103
RNAaseH, 104
rRNA, 5, 7

S1 mapping, 57
Saccharomyces cerevisiae, 119
Sanger method, 48
*Sau*3A, 36
Selectable marker, 82

Sendai virus, promoting membrane
    fusion, 33
Shine-Dalgarno sequence, 10
Signal peptide, 8
Signal sequences, 15
Single strand conformation
    polymorphism (SSCP), 71
Site-directed mutagenesis, 88
*Sma*I, 36
Small-eye mutation, 138
Solenoid, 19
Somatic cell hybrids, 33
Southern blot, 23, 27, 33, 38
Sp1, 15
Sp1 site, 16
Splice junctions, 59
Splicing, 17
    differential, 17
Splotch gene, 138
SSCP, 71
Stable transformants, 127
Staggered termini, 37
Stem-loop structure, 13
Sticky ends, 37
Subtraction hybridization, 108,
    111
Subtraction library, 108
SV40, 2, 127

T-cell receptor, 108
Tandem repeats, and unequal
    crossing-over, 42
Taq polymerase, 60, 61
TATA box, 15
Termination codon, 6
TetOFF, 129
TetON, 129, 130
Tetracycline, 129
Tetracycline operon, 129
Tetracycline repressor, 129
TFIID, 15
Thymidine, 1
Thymine, 1
Tm, 23–25, 61
Topoisomerase I, 88

Transcript mapping, by nuclease
    protection, 56
Transcription, 3
    start site, 59, 63
    termination, 16
Transcription factor binding site, 59
Transcription factors, 15, 16
Transcriptional activation domain,
    124, 129
Transcriptional control, 10
Transferrin receptor mRNA, 13
Transformation, of eukaryotic
    cells, 119
Transgenic mice, 131
Transient expression, 127
Translation, 6
    in vitro, 84
    rate of, 8
    start site, 59
Trinucleotide expansion, in
    DMPK gene, 40
tRNA, 5, 6
Tumor typing, 140
Two-hybrid system, 124, 125

Unclonable sequences, 98
Uracil, 3

Variable number tandem repeat
    (VNTR), 41
Vector, 81
    BAC, 95
    bacteriophage, 82
    cosmid, 94
    extrachromosomal, 128
    insertion, 91
    integration of, 120
    PAC, 95
    plasmid, 82
    recircularization, 87
    replacement, 91
    retroviral, integration, 129
    retroviral, 128
    shuttle, 120
    yeast, extrachromosomal, 122

VNTRs, 41, 44, 65
VP16, 129

Waardenburg's syndrome (WS1), 138
Wobble base, 6
WS1, 138

YACs, 123
Yeast artificial chromosomes
    (YACs), 123